HOW TO LIVE ON MARS

About the Author

A man whose career has required the use of several names, **Robert Zubrin** was born in New Plymouth in 2071 and graduated Heinlein High in 2099. Due to an unfortunate accident that caused his parents' payoff to the school administration to be misplaced, he was mistakenly ranked near the bottom of his class and was forced to accept employment from NASA for seven years (a time span he calls his dark period). Eventually, however, he was freed, and finding honest work, achieved interplanetary renown and financial success through a series of highly lucrative ventures in the areas of prospecting-claim evaluation, areopaleontology, and preterraforming real-estate development. He is no (proven) relation whatsoever to his twentieth-century namesake, a humorless astronautical engineer who developed the Mars Direct mission plan, authored the classic treatise *The Case for Mars*, and led the founding of the Mars Society in 1998. He regrets any confusion his current nom de plume may have caused.

BOOKS BY THE OTHER ROBERT ZUBRIN

The Case for Mars: The Plan to Settle the Red Planet and Why We Must

Entering Space: Creating a Spacefaring Civilization

First Landing

Mars on Earth: Adventures of Space Explorers in the High Arctic

The Holy Land

Benedict Arnold: A Drama of the American Revolution in Five Acts

Energy Victory: Winning the War on Terror by Breaking Free of Oil

Robert Zubrin

HOW TO LIVE ON MARS

A Trusty Guidebook to Surviving and Thriving on the Red Planet

THREE RIVERS PRESS • NEW YORK

This is a work of fiction. Names, characters, places, and incidents either are the product of the author's imagination or are used fictitiously. Any resemblance to actual persons, living or dead, events, or locales is entirely coincidental.

Published in the United States by Three Rivers Press, an imprint of the Crown Publishing Group, a division of Random House, Inc., New York.
www.crownpublishing.com

Library of Congress Cataloging-in-Publication Data
Zubrin, Robert.
How to live on Mars : a trusty guidebook to surviving and thriving on the red planet / Robert Zubrin.—1st ed.
p. cm.
1. Mars (Planet)—Humor. I. Title.
PN6231.M32Z83 2008
813'.4—dc22

ISBN 978-0-307-40718-4

Printed in the United States of America

10 9 8 7 6 5 4 3 2 1

First Edition

To Jamie Caitlin West Lutton:

Sage, Wit, and Muse;

Proprietor of Seattle's Twice Sold Tales, *the greatest, wildest, and weirdest used bookstore in the world;*

Spreader of light and laughter;

Original character;

And the truest of friends;

This book is cheerfully dedicated.

Contents

List of Illustrations

Preface

So you've made the decision to break the surly bounds of Earth and head on out to the Martian frontier. Good move! Mars is where the future is. Its wide-open spaces are waiting for folks like you with guts and gumption to go out and make your mark. It's a new world, ready for a young civilization to be born, and rife with history raring to be made. Now that you have signed up, you can be one of those on the make.

Yet you mustn't kid yourself. Mars is a great place, but it's no picnic. Many people have gone to Mars with high hopes and dreams of making it big, only to fall flat on their faces. The Red Planet can be very harsh to those who come unprepared. You've sold your house and liquidated all your savings for a one-way ticket; do you really want to risk ending up dead broke, forced to accept a seven-year contract cleaning sewage recyclers, spending your nights hot-bunking with two other guys who stink as bad as—or God forbid, even worse than—you do? No? I thought not.

And that's the least that can happen to you. With an unbreathable atmosphere less than 1 percent as thick as Earth's, and nighttime temperatures that fall to −90°C, Mars is loaded with perils for the careless or unwary. Make one slip on Mars and you can wind up very, very, very dead.

Fortunately, there's no reason to worry. Because now, thanks to many contributions from the smartest and most battle-tested Mars pioneers from every walk of life (including yours truly, a person who has been working on Mars for so long that people sometimes confuse me with my twentieth-century namesake), a treasury of priceless experience has been compiled to offer you everything you need to know not only to survive, but to thrive—indeed, to *succeed beyond your wildest dreams* in everything you do on the Red Planet.

Look no further; this book has all the answers. All you have to do is study and absorb its kernels of wisdom. After reading *How to Live on Mars,* you will prevail over any obstacle that Mars presents you.

In this invaluable compendium, you will learn every trick of the successful Martian pioneer, including:

- How to get to Mars.
- How to choose a hab. (Remember three things: location, location, location.)
- How to choose a life-support system. Which is the right recycling system for you? (Bioregenerative or physical-chemical? Going green may be stylish, but the smell is not for everyone.)
- How to choose your first ground rover, including tips on how to spot a lemon. (Hint: Don't buy anything developed for the Moon program.)
- How to choose a spacesuit. Not all spacesuits are alike! You need a suit that not only works well but feels good, looks smart, and won't go out of style. The old Earth saying goes:

"Clothes make the man." The same is true on Mars. Your spacesuit doesn't just keep you alive, it also defines your image!

- How to stay alive in the desert. Everyone gets stranded sooner or later, but only the prepared survive. There's much more to rocks, sand, and dust than meets the eye!
- The cheapest way to provide radiation protection for your home. Yes, good shielding is important, but that's no reason to be taken for a ride by some snobbish salesman pushing designer brands. You can make do for less.
- How to snag your first job, including priceless advice on which jobs to seek and which to avoid.
- What ventures to invest your savings in, including a list of the ten best high-tech start-ups and real-estate deals on Mars.
- Which plants to grow in your greenhouse, and the best techniques for producing crops that are actually edible. (Hint: Don't use *your* greenhouse for human waste recycling.)
- How to save money by making your own rocket fuel and blasting explosives at home.
- How to make synthetic alcohol for fun and profit.
- How to make your own steel, glass, plastic—or anything else that you want—out of nothing but dirt and air.
- The best places to shop for gear, including tips on where to find the top bargains, as well as the key scams to avoid.
- How to choose a prospecting partner or an exploration team.
- How to make discoveries that will make you rich and famous.

- How to profit from the terraforming program. Make global warming your friend!
- The best places to go for meeting members of the opposite sex, and the most effective pickup lines to use. This is important. Mars is not Earth—things are done differently here. This section alone is worth many times the price of the book.
- The best places on Mars to have some real fun. Don't miss out!
- How to choose a spouse.
- How to choose the best school for your kids.
- How to avoid space-agency import fees, tariffs, taxes, utility charges, and other unnecessary expenses.
- How to instantly spot NASA or ESA inspectors, regulators, revenuers, and shrinks.
- How to wage a successful legal defense against bureaucratic persecution, regardless of the evidence.
- How to win elections. Public office can be profitable too!
- And a thousand other priceless tips.

So now the book is in your hands, and with it, the key to your future. You can put it down and end up an indentured sump cleaner, or you can make the wise choice, become informed, and rapidly rise to riches and universal esteem as one of the ranking citizens of our great new civilization. The choice is yours!

To the frontier the American intellect owes its striking characteristics. That coarseness of strength combined with acuteness and inquisitiveness; that practical, inventive turn of mind, quick to find expedients; that masterful grasp of material things, lacking in the artistic but powerful to effect great ends; that restless nervous energy; that dominant individualism, working for good and evil, and withal that buoyancy and exuberance that comes from freedom—these are the traits of the frontier, or traits called out elsewhere because of the existence of the frontier. . . . At the frontier, the bonds of custom are broken, and unrestraint is triumphant.

—Frederick Jackson Turner, *The Significance of the Frontier in American History*, 1893

When a place gets crowded enough to require IDs, social collapse is not far away. It is time to go elsewhere. The best thing about space travel is that it made it possible to go elsewhere.

—Robert A. Heinlein

Hell, there are no rules here—we're trying to accomplish something.

—Thomas A. Edison

The Basics of Survival

[1]

How to Get to Mars

Anyone can get to Mars. The key thing, however, is to get there alive and without losing your shirt. I take it we are in agreement on this point. In that case, I have advice for you: Don't take a cycler. Don't even think about it.

Cycling Spacecraft

Yes, I know what you've heard about those wonderful cycling space stations—"castles in the sky," with accommodations far roomier than any one-shot spaceship could ever hope to offer, orbiting eternally back and forth between Mars and Earth. Well, that's just the problem. These stations have been orbiting forever—or ever since 2042, anyway—back and forth, back and forth, and consequently they are now so full of green and brown microbial slime that the stench is almost enough to kill you long before the space-mutated microbes that infest their broken-down water recycling systems even have a chance to turn your insides into

The Russian space station Mir, which had to be abandoned and destroyed in 2000 after fourteen years of orbital use filled her with green slime. The cyclers have been in space much longer.

bloody mush. I'm sure these stations must have sounded like a great idea back at the beginning of the century: "Build them big, launch them once, and reuse them forever."

But the reality is that these fabled flying castles are nothing but ancient, unsanitary, filth-filled bilge cans, designed by a bunch of overpaid, underworked government clowns who most likely were born when George Bush was president, and whose engineering competence reflects that fact.

Perhaps you feel I exaggerate? Well, then, ask yourself this question: Have you ever met anyone who took a cycler to Mars and also chose to take one back? No? I didn't think so.

And, in case you are thinking, "But it's worth living in a flying latrine for eight months in order to save a bundle on the interplanetary transfer," think again. Yes, it's true that a berth on a cycler can be had for a song, but they take you to the cleaners for your seat in the taxi capsule that shoots you out from Low Earth orbit to perform Hyperbolic Rendezvous (HR) in interplanetary space with the cycler as it whizzes by Earth. And that's just the beginning. Because you'll also need to buy HR insurance to cover the sky-high cost of capsule rations if your taxi misses the cycler—which it does half the time—and you are forced to live on synthetic pemmican and recycled metabolic products for two years until your sardine-can (which is on a trans-Mars trajectory, after all) cycles back to Earth and you can try again—assuming you are still alive, which you probably won't be, because the taxi capsules lack proper solar-flare storm shelters and in the course of two years you are bound to get zapped.

But even if you successfully make it to the cycler, it's just not worth it when you consider the fact that you'll be stuck for the duration of your trip with a bunch of other Joes who are cool with the idea of cesspool-economy spaceflight. In fact, you'll be stuck with them for a lot longer than that, because no Martian who still has his, or especially, her, sense of smell will be caught dead hanging out with cycler scum. Think about it. You are going to Mars to begin a new life. Do you really want to start out, and stay, at the bottom of the barrel? Do you hope to meet your soul mate, or just find some action? Either way, if you are dumb enough to go to Mars by cycler, the polar caps will melt before you ever get a date.

Nuclear-Electric Propulsion

Now that we have ruled out the cyclers, how *should* you go? If you are rich but not terribly astute, you are probably already thinking that splurging a few million shekels for a ticket on a fast nuclear-electric ion-drive spaceship might be a better choice. That's certainly the choice that NASA would prefer you take, because they're still trying to justify their latest boondoggle. I'm sure you've seen their vids: "Nuclear-Electric: modern propulsion, taking you anywhere, anytime."

Well, it's your money, and if you're willing to burn it, that's your business. But before you do, there's something you ought to know: Nukey ships are for the birds. Yes, it's true that in theory they can really work themselves up to incredible velocities, but they take over a year to do so. So you actually get to Mars quicker on a cycler (to say nothing of doing it by a six-month Type 1 ballistic transfer, as the first explorers did, way back when). And while those fancy techno-wonders certainly look majestic with their vast glowing arrays of radiators, power conditioners, ion thrusters, and the rest, all that gear is vulnerable to breaking down. If it chooses to do so after the ship has gotten up to speed but before she can slow herself down, you'll find yourself bound for a random destination somewhere well beyond the Oort Cloud—a predicament that gives new meaning to the NASA slogan "Taking you anywhere, anytime." Of course, the skippers of such Flying Dutchmen don't fancy such an outcome any more than you do, which is why they have all that ultrafine print in your transit contract that allows them to order *you* to do a little extravehicular activity (i.e., "go EVA") for in-flight propulsion-system repairs near the ship's reactor.

Please don't get me wrong. You're coming to Mars, so naturally you're going to pick up a few dozen Rems here or there. That's OK. A little bit of radiation never hurt anyone. It's a *lot* of radiation that you have to worry about. And unfortunately, those nukeys are as hot as pistols. That's why they have that 200-meter-long boom between the reactor and the crew habitat—it drops the dose rate by the square of the distance. But the power generation and conditioning gear your loving ship's officers will send you out to fix is all within 5 or 10 meters of the core. Yes, it's behind the reactor's shadow

A nukey ship. She looks cool, but note that her engines are firing the wrong way to slow down for Mars arrival. Farewell, voyagers.

shield, but don't let them fool you. Most of the hab's shielding actually comes from the row of propellant tanks strung out along the boom, and you'll be going forward of all that stuff. Furthermore, not only will you be exposed to at least a ten-thousandfold increase in dose rate from the reactor, you will be working directly on equipment that has absorbed a lot of neutrons over time, and so will be medium-level radioactive itself. Are you starting to get the picture?

So, if the two mainstream choices of a cheapo stinky cycler or a wildly overpriced nukey ride to glory are both unappealing, what *is* the answer? How can you get to Mars alive, unstunk, undosed, and unbroke? Believe it or not, there is a way.

The secret is to ride the freight. That's right, *take the freight*. Of course, not just any freight will do. If you want to get to Mars with your lungs still pumping, you need to ride aboard freight that includes life support. What freight is that? It's the new hab modules being shipped out for use on the Martian surface.

Cargo Flights

This is one of the best-kept secrets in the whole Mars program. NASA doesn't want you to know about it, because they want you to shell out your cash to buy a berth in one of their nukey ships, or at least spring for a ticket on a decrepit cycler. But there is no point to doing either. Every year, dozens of brand-new, spiffy-clean surface habs are fired off to Mars to provide housing for the more successful members of the colony's growing population. There's no reason you can't get to Mars aboard one of those. It's the best deal in the solar system. You can ride to Mars for free, or, if you have the right references, skills, and attitude, you can even *get paid* by the hab's buyers to house-sit their prospective property for them during the outbound flight. Why should the new owners pay you to ride to Mars in their house? Simple: Nobody wants to buy a lemon, and your surviving the trip provides them with an unfakable proof of product quality.

Don't worry. Those NASA Safety Office "Don't Be a Guinea Pig" vids warning people against riding hab freight are nothing but pure bull. Think about it. If you arrive dead, the hab sellers don't get paid, so you can bet your bottom dollar that they will do *everything* they can to make sure you get there in the pink. Ask for first-class rations—you'll get them! Insist upon a good old-fashioned, quick, Type 1 flight plan—they'll agree! You can even demand they add a tether and spin the thing around its spent trans-Mars injection stage to give you artificial gravity. They'll do it! That's right, they'll do it for *you* because they *care* about you making it to Mars alive and healthy.

And there's another bonus to flying by private freight—you can take along some choice cargo of your own. There's plenty of little special items in *very* high demand on Mars that the well-meaning NASA nannies somehow just can't bring themselves to include in the official supply manifest. You can help solve this problem and do well by doing good. What items should you include in your off-the-manifest stash? That depends. I suggest you consult with the hab's buyers. Trust me, they didn't get rich on Mars by being dumb. Let them be your teachers. Work with them. Be a team player. Show the smart set that you're the sort who can help create some action, and don't worry, they'll make sure you get your piece. That's the Martian way.

Technical Note (WARNING: High Science Content) Interplanetary Orbits

If you want to travel from Earth to Mars, you need to choose an appropriate orbital path.

The classic choice for an interplanetary travel orbit is known as the Hohmann Transfer, named after a German mathematician who discovered it in 1925. The Hohmann transfer is an ellipse, one of whose foci is at the sun, and whose perimeter is tangent to the orbit of Earth on the short side of its major axis (the position closest to the sun or "perihelion") and tangent to the orbit of Mars on the other (the position farthest from the sun, or "aphelion"). So if you take a Hohmann transfer, you will travel 180 degrees around the sun, leaving Earth in, for example, the six-o'clock position to arrive at Mars in the opposite twelve-o'clock position. The Hohmann transfer is thus the longest direct route between Earth and Mars, but despite that, it is preferred by many mission planners because it has the smallest propulsion requirements. For this reason, the Hohmann transfer is also known as the minimum-energy orbit. It may not be the fastest, but it is the cheapest and easiest path between planets. When employed between Earth and Mars, it dictates a one-way trip time of eight and a half months.

The Hohmann transfer is a special case of a general type known as conjunction-class orbits. Ancient astrologers, with their geocentric worldview, deemed Mars to be in "opposition" to the sun when the two were on opposite sides of Earth, and in "conjunction" with the sun when the Red Planet was on the same side. This latter condition occurs when Mars is behind the sun as seen from Earth, which is the case with the targeted position of Mars at the moment of launch of a Hohmann transfer. Twentieth-century spaceflight engineers at NASA's Jet Propulsion Laboratory, apparently deeply steeped in astrological lore, chose to call the Hohmann transfer and similar orbits "conjunction," and the name has stuck.

As implied above, the Hohmann transfer is not the only conjunction orbit. If you pour on some extra departure propulsion, the transfer orbit can be enlarged so that its aphelion is beyond the orbit of Mars. In this case, the transfer orbit will not be tangent to that of Mars, but rather intersect it in two positions, one on the outbound and the other on the return. So, for example, if you leave Earth at the six-o'clock position relative to the sun, you will now be able to reach Mars, not at the twelve-o'clock position, but at either the one-o'clock or eleven-o'clock positions. If your point of view is such that the planets and the spacecraft are all traveling counterclockwise, then the trip from six o'clock directly to one o'clock will be shorter than the Hohmann transfer; while the trip from six o'clock out past Mars to hit the Red Planet on the way back at eleven o'clock will take longer. Both of these routes are still considered conjunction orbits, but the first, quicker path, is known as a Type 1 conjunction, while the second is Type 2.

Compared to a Hohmann Transfer, a Type 1 conjunction has the disadvantage of requiring somewhat more fuel, but it offers the benefit of a shorter trip time, say six months one-way instead of eight and a half. It is thus the preferred choice for human interplanetary travel. The Type 2 conjunction, on the other hand, also requires more propellant than the Hohmann Transfer, but actually lengthens the trip, perhaps to eleven months.

So, here's a hint: If you're short on cash, take the Hohmann Transfer. It's the cheapest way to fly. If you've got some extra krill to spend, or are working for someone who does, go Type 1. But don't let yourself get stuck on a Type 2, as you will end up paying more for a longer trip. Sometimes ships have to go Type 2 when their launch schedule requires it, but this is something that you should really seek to avoid.

This returns us to the issue of the cyclers. A long time ago, somebody had the bright idea that it would make sense to launch one or more large space stations in permanent transfer orbits between Earth and Mars. Once established, such large habitable structures would not have to be launched again, and people could make the trip just by catching the cyclers in small, fast "taxi" capsules fired out from either Earth or Mars orbit as the cycler passed by. Aside from the risk of missing the connection (you only get one shot), the idea sounds good, but it introduces many complications. For example, the one-way Hohmann transfer travel time is eight and a half months, so its round trip time is seventeen months. This does not synchronize with the twelve-month orbital period of Earth. Thus, if a spacecraft were to depart Earth on a Hohmann transfer and allowed to orbit back again, when it returned the Earth would not be there to meet it. For this reason, cyclers cannot use Hohmann transfers. A cycler could make use of a higher-energy orbit that goes out beyond Mars to return to Earth twenty-four months after departure, and thus effect a rendezvous with the home planet. In such a case, however, the cycler's itinerary between

Earth and Mars would involve a short Type 1 leg as well as a long Type 2 leg. These could be assigned to either the outbound or inbound routes, according to preference, but passengers going one way or the other would always end up getting stuck with the slow boat. Furthermore, while a cycler with a two-year period could be counted upon to rendezvous with Earth every orbit, it would not be synchronous with Mars, whose orbital period is twenty-three months. Thus, in order to make such a transportation system effective, there would have to be a number of cyclers, each one returning to the Earth's orbit at a different point and only being used for Earth-Mars transportation once every six or seven years.

While all orbits require more propulsion than a Hohmann transfer, the additional acceleration required to attain a medium-energy Type 1 (or Type 2) conjunction orbit is relatively modest (so long as you don't try to push the one-way travel time below five or six months). If you really want to get to Mars fast, say in three months or less, you need to use much higher-energy orbits. Such orbits are not practical choices for spacecraft dependent upon chemical propulsion (with exhaust velocities of about 4.5 km/s) or even nuclear-thermal propulsion (which can have exhaust velocities of 9 km/s). Instead, very advanced propulsion systems, such as nuclear-electric ion drives (with exhaust velocities of 50 km/s or more) are required. The problem with such ultrahigh-energy orbits, however, is that their maximum speed exceeds the escape velocity of the solar system! So, if the propulsion system should fail after acceleration but before deceleration, the vessel will find itself on a one-way trip to infinity and beyond.

A final orbit type worth mentioning as a footnote is the opposition-class trajectory. In an opposition mission, the trip to Mars is divided into two unequal legs. One of the legs can be a

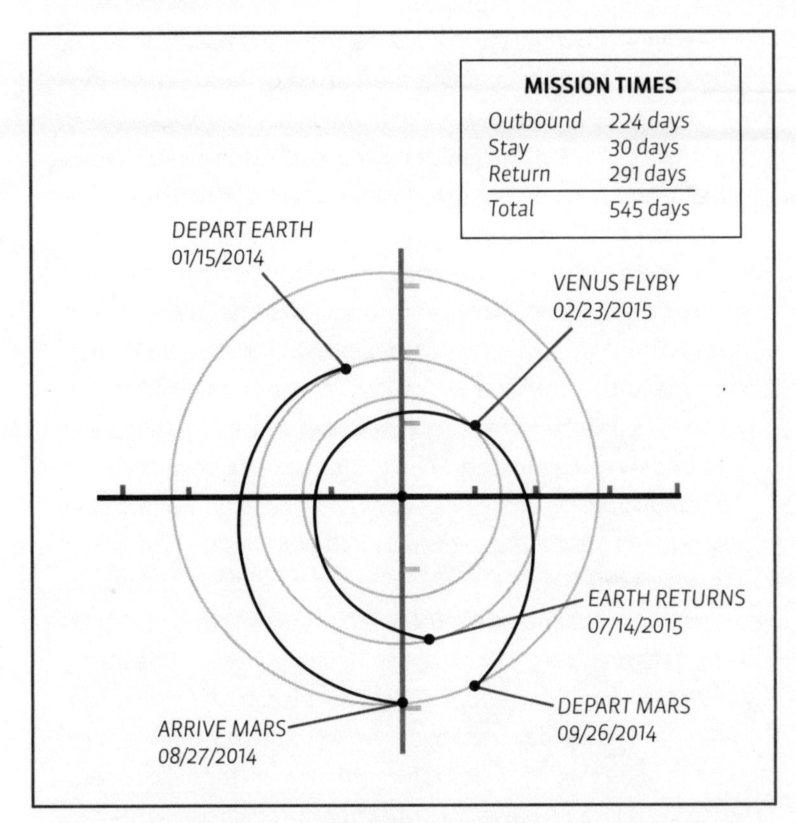

Diagram taken from an ancient NASA document, showing plans for an opposition-class mission to Mars. Note the long transit times, the absurdly short stay time, and the toasty Venus "fry-by" the crew would have experienced on the return flight. What were those guys smoking?

conjunction transfer, but the other is an orbit that cuts through the inner solar system—typically as close to the Sun as Venus—in order to swing the ship around and make it possible for a transfer to occur while Mars is on the opposite side of the Earth from the Sun (and thus named "opposition," by the old geocentric JPL astrologers).

The opposition trajectory involves more propulsion than any reasonable conjunction path, and one of its two travel legs is much longer, typically requiring as much as fourteen or fifteen months. In addition, the radiation and heat exposure of the passengers and crew during the close-in solar pass (known to the experienced as a Venus fry-by) can be quite nasty. In short, it's a really dumb flight plan, which is why it is almost never still used. It was dreamed up by NASA planners back in the twentieth century, because while it does not minimize flight time, it can minimize round-trip time of a Mars mission—since the need to stay on Mars to await a (smart) conjunction-class return launch window to open is eliminated. For the old-time Mars explorers, flying conjunction meant taking a six-month Type 1 trajectory each way, with an eighteen-month stay on the Red Planet in between. In contrast, the opposition plan allowed a mission with one six-month flight leg and one fourteen-month leg, with only a one-month stay on Mars. But since the idea of going to Mars was to explore it, which takes time on the surface of the planet, this was a pretty silly idea even back then. The countervailing benefit that the opposition trajectory minimized time away from home seems a quite absurd justification; after all, if your primary criterion for a good mission is minimizing time away from home, the obvious solution is simply to never leave!

In any case, you're going to Mars to stay. So fly conjunction.

[2]

How to Choose a Spacesuit

If you have followed my advice thus far, you've now arrived on Mars healthy and happy, with some good connections and a nice little nest egg to help you get started. Well done. Nevertheless, there are many more critical decisions you will have to make if you wish to succeed on Mars. The first of these involves how to dress.

There's an old Earthling adage that "Clothes make the man," and this is emphatically true on Mars. While of some importance, I won't waste time discussing issues of indoor attire, as the guiding rules in that arena are not very different from those that apply on Earth. However, as the most lucrative opportunities on Mars involve outdoor activity, you are very quickly going to have to face the matter of choosing a proper spacesuit.

I assume you had enough brains not to buy one on Earth, and thus avoided paying retail launch and shipping costs. But now you need to know how to shop for a good suit on Mars.

Make no mistake: This is a critical issue, and one for which most new immigrants come woefully unprepared. You need a spacesuit that not only looks good but works well, feels good, won't go out of style, and most important, won't brand you publicly as a total newbie ripe for the plucking. Never forget: Your spacesuit defines your image!

Spacesuit Pressurization Systems

There are two kinds of spacesuits: pneumatic and elastic. Pneumatic spacesuits are the older type that use gas to pressurize your body to at least 3 pounds per square inch (psi), thereby preventing your blood from boiling when you go outside in space or in a nearly spacelike environment such as currently prevails outdoors on Mars. These are the kind of suits that the Apollo astronauts wore on the moon and the crew of the *Beagle* employed on the first mission to Mars. They are puffy, bulky, and clumsy, but nevertheless, still in use. In contrast, the more modern suits use skin-tight elastic fabric to pressurize your body without the use of inflation gases. They are sleek, stylish, highly flexible, and if you have a good body, they do a great job of showing it off. When you go to the S&R store suit department, these are the outfits that all those ultrasexy sales models in the front section will be sporting.

In his journey home from Troy, Odysseus had to sail his ship safely past the rocks upon which the Sirens were sweetly singing their treacherous song. Well, a lot may have changed since then, but some things are still the same: If you want to survive your Martian odyssey, you're going to have to sail past the Sirens too. I don't care what it takes. If wax earplugs will help, use them. If you think you need to have some friends tie you to a mast and carry you past, do that. But under no circumstances must you allow S&R's Sirens to sell you an elastic suit.

Yes, I know *they* look smashing in those getups, and maybe you think you will, too. Or maybe you just want to get on their good side by buying what they are selling. Don't do it. No matter how much they encourage you, no matter how much they mock the look of the alternative "puff-boy" pneumatics, don't listen. Believe me, if you do, you'll be the butt of jokes from every man and woman you meet from that moment on until you either get rid of the outfit or flee the planet.

Elastic suits are for idiots. In the first place, unless you have the body of a supermodel, they really don't look that good. (If you actually do, S&R will be glad to give you one on the house for promotional purposes.) I mean, guys, really, do you want everyone to see every time you have an interesting thought? And when you do, and your thingy starts pushing out the elastic in that special place, are you prepared to deal with the consequences if the heating elements embedded in that section of fabric don't respond quickly or accurately to deliver protection to the parts in question against the surrounding –50°C cold? Look, this is not a new problem. Men's clothing had a similar feature in old Europe during the Renaissance, and there is a reason why it went out of style.

Elastic spacesuits can be very stylish, but only if you have the right figure. Otherwise, they can be embarrassing. Most people look better in a pressurized classic.

But that's just the beginning. Even if you are prepared to cope with this delicate issue (or if you are a gal and don't have to), there is another, more basic, problem. Elastic suits are made to fit you *exactly* at the moment of purchase. If you gain any weight or lose any weight, they are no good. Those skinsuit girls have to eat like birds in order to stay within the required parameters. Are you prepared to live like that?

In short, the answer is to go to the back section of the store and choose a pneumatic suit off the rack. Yes, they are old-fashioned, but their look is classic. They've been in use now for over a hundred years and they haven't gone out of style. Think of yourself as another Neil Armstrong or Becky Sherman, because you'll be wearing what *they* wore when they made history. And if they didn't look sleek and sexy, so what? They looked like who they were—folks with the Right Stuff. And that's the look you need to project as well.

Having made the key decision to go pneumatic, there are several more issues to cover. First, I strongly advise you to buy new, not used. You've saved a lot of dough by going pneumatic instead of elastic, so unless you are absolutely dead-broke, you don't want to pinch pennies by trying to reuse someone else's pneumasuit. There are things that are worth more than money, and having a suit of your own is one of them. Yes, I know the used dealers claim to have triple-washed the things, and in some cases they actually have, but it doesn't matter. The smell just doesn't come out. Wearing a previously occupied pneumasuit is like traveling in a cycler, an experience to be avoided at all costs.

Second, while I am totally opposed to the concept of elastic spacesuits, there are new models of pneumatic suits that feature elastic *gloves*, and these *do* have some merit. If you are going to be doing outdoor work that requires special dexterity and you can afford the extra cost, you might wish to consider a pair. Keep in mind, however, that you will still need to wear thick cold-weather gloves over these or you will freeze your fingers off the first time you touch anything.

Finally, there is the matter of color. For whatever reason, the old-time NASA astronauts wore white, and in keeping with the desire to project a heroic image, you might choose to do so as well. I, however, advise against it. Red dust is everywhere on Mars and if you choose to go with classic white, your suit is going to look like shit in no time. So go with Mars tones instead. I know it's a bit of a compromise, but there is really no other sensible choice.

Oxygen Supply System

We now need to discuss issues associated with the suit systems. First and foremost among these is the oxygen supply. Here you will be offered two choices: compressed gas or cryogenic liquid. I recommend the latter. While it is true that all cryogenic oxygen systems are susceptible to slow losses through boiloff, you can store so much more liquid oxygen in your tank than gas that this doesn't matter. The compressed gas systems have a maximum endurance of twelve hours; with cryogenic liquid systems you can get up to thirty-six hours. This difference can save your life if things don't go as planned out there on the planitia. On the other hand, compressed oxygen is much cheaper. But that's only if you are *buying* it. Almost all the ground rovers in use on Mars use liquid oxygen in their propulsion systems, so if you are out on a job, you can always get liquid oxygen for free just by siphoning it off one of the company vehicles when the boss isn't looking. So, in reality, the liquid oxygen systems are not only safer, but more economical to operate as well.

Spacesuit Power System

After oxygen supply, the next most important system in your suit is power. If you can afford it, I recommend methanol/oxygen fuel cells. These are quite reliable, and, if you've taken my advice and selected cryogenic liquid oxygen for your breathing system, you only need to carry a modest amount of methanol to give you a terrific power reserve. However, because they are so good, the methanol/oxygen fuel-cell power systems are in great demand and go for top dollar. As a new immigrant, you might not be able to afford one. If that is your situation, I recommend that you choose batteries. That's right, ancient, prehistoric *batteries*. Under no circumstances should you fall for the NASA line and buy one of their new hydrogen/oxygen fuel-cell units.

Look, I've heard it all, and these people are crazy. It's just like their nukey ships; they're always trying to make you use something they've developed in order to validate their activity. It's not that they don't do their engineering right. It's that they never really think about whether they are engineering the *right thing*. So here they come again with their new H_2/O_2 Technology Space Suit Fuel Cell Unit (HOTSS-FuCU) systems. Yes, it's true that these units are quite efficient, and the liquid hydrogen only weighs one-eighth as much as the 2 kg of methanol it replaces. Big deal. It saves you the trouble of carrying 1.75 kg of fuel, which only weighs 0.65 kg on Mars anyway. But you're stuck having to mess around with 20 Kelvin hard cryogenic liquid hydrogen, which is always boiling away.

"But the suit's liquid oxygen is boiling away too," they say, "so what's the problem?" That's *exactly* the problem! Mars to Houston, hello? Did any of those nitwits ever stop to think what might happen if someone wearing one of their HOTSS-FuCU things ever got stuck in a hab airlock during an unscheduled operations hold? Well, let me spell it out for you: There you are, standing in the crowded lock with a bunch of other suited-up folks, wondering when the hold will end. Your hydrogen starts to boil off. It mixes with the oxygen boiloff accumulating in the lock chamber. Somebody rubs his boot on the floor. There is a tiny spark. Kablam! No more airlock, no more hab, and no more you. So much for HOTSS-FuCU. As I said, if you can't afford methanol fuel cells, use batteries.

Communication System

The next spacesuit system we need to discuss is com. When you are out on EVA, you need a way to talk with other people, and in the thin air of Mars, sound won't carry very far. Furthermore, even when you are quite close by, the fact that both you and your interlocutor are walled off from each other by your respective spacesuit helmets can make sound conversations difficult. So you will need a radio, which is why all Martian spacesuits are equipped with a standard 2-meter (144 MHz) VHF rig. That much is mandatory. But you still need to make the right choice of control system for your suit's radio transmitter.

Most suit radios come with factory-installed, voice-activated microphones, and the S&R salespeople will tell you that's all you need. Wrong. Voice-activated mikes are all right as far as they go, but you need the option to turn them off and switch to manual push-to-talk mode. I know the idea of using push-to-talk radios in this day and age sounds hopelessly archaic, but trust me, it's essential. If you can't switch to push-to-talk, you will be left at the mercy of your suit's radio computer to decide when you are talking and when you are not—or to be more precise, when you can't talk and when you must. If you cough, chew gum, or fart in your suit, the sound will be transmitted. If you grunt, hum, burp, or curse under your breath, it will be transmitted. If you start breathing hard, your panting will be transmitted, and sometimes your respiratory sounds will be transmitted even if you are not breathing particularly hard because your radio computer has decided on its own to up its gain a few hundred decibels. So there you are, with your exhalations blasted directly into the helmetphones of every other tyro on the planitia, thereby activating their suit mikes to feed the broadcast right back into your helmet radio, which amplifies it again and again—until everybody goes deaf from the uncontrollable feedback.

Of course, if you are out working only with experienced Mars hands, who all use push-to-talk radios, there might not be any feedback catastrophe, but everyone else's transmissions will still be jammed by your respiratory solo, and upon determining the source of the problem, the rest of the gang will come looking for you.

Now this problem has been with us for some time and, as I'm sure you will be told, NASA recently came up with a (pricey) high-tech solution for it: their "think-to-talk" (TTT) system. These remarkable gadgets work by sensing neural impulses inside of your brain, analyzing them, and transmitting their content as words. Their development is an incredible technical accomplishment—but what can I say? TTT takes voice-activated suit radio issues to a whole new level. Instead of uncontrolled broadcasts of your panting, humming, and gum-chewing, with TTT you will now be in a position to effortlessly treat all of your coworkers to a piece of your mind—with the "piece" in question to be chosen by your friendly TTT nanoprocessor. At the product rollout, they demonstrated the first unit on the NASA administrator, and the Family News Service had to cancel its coverage.

Another feature that is built into every suit radio is the MPS receiver and locator beacon. This takes a fix on your position by reading the signals from the Mars Positioning System satellite constellation, and then broadcasts that location along with your suit radio's registration number once per minute. This system can be quite useful in helping to assure your safety, as it allows your movements to be tracked by the settlement central computer at all times. By the same token, however, it can be very inconvenient if you want to engage discreetly in many highly rewarding independent business ventures, or even join in private social activities following the true imperatives of your heart. Accordingly, you need to make sure that your suit locator beacon features the dual-data-stream option, which gives the user the choice of providing it with actual positional data or an alternative travel log, as the occasion requires. Such upgraded locator beacons may cost a little more, but they are well worth the price.

Thermal Control and Waste Management

Your suit's thermal-control system is equally important. You'll want one that includes a dual-redundant helmet defogging system, with *both* internal warm gas blowers and electrical heating wires embedded right in the transparent dome. Otherwise, you can easily be blinded when its inside frosts over.

Another feature you should insist upon is an electrical heating system installed within the soles of your boots. The Martian ground temperature is generally above freezing during the daytime, but at night it falls to –90°C. They say that way back during the first mission, ol' Becky Sherman actually ran 200 meters during the middle of the night from the *Beagle* to her Earth Return Vehicle in unheated boots. I don't know how she did it, but I sure know why she *ran*. Believe me, it's not something you want to try. Sometimes sadistic S&R models will try to get revenge against newbies who choose a pneumasuit by offering them electro-heated socks as an alternative to heated boots. Don't fall for that one. Yes, it's true that electrosocks save power compared to boot warmers, because they are closer to the skin and don't waste as much heat to the environment. The problem is, when human sweat seeps in among the socks' heating wires, it can short them out, and sooner or later the electro*socks* start delivering electro*shocks*. (This is another reason why I find elastic skinsuits unappealing—they include live heating wires in contact with not just the feet, but *every* part of the body. So if you do buy a skinsuit, *be careful not to sweat.*)

Finally, there is the issue of the suit's Waste Entrainment Temporary Cachement System (WETeCS), otherwise known as its diaper. Here, the answer is simple: Choose maximum capacity. When you've got to go, you've got to go, and if you are in a suit, that's where you are going to go. WETeCS overflows are really bad news. So play it safe. Choose a max cap WETeCS for your suit. Trust me; you'll be happy that you did.

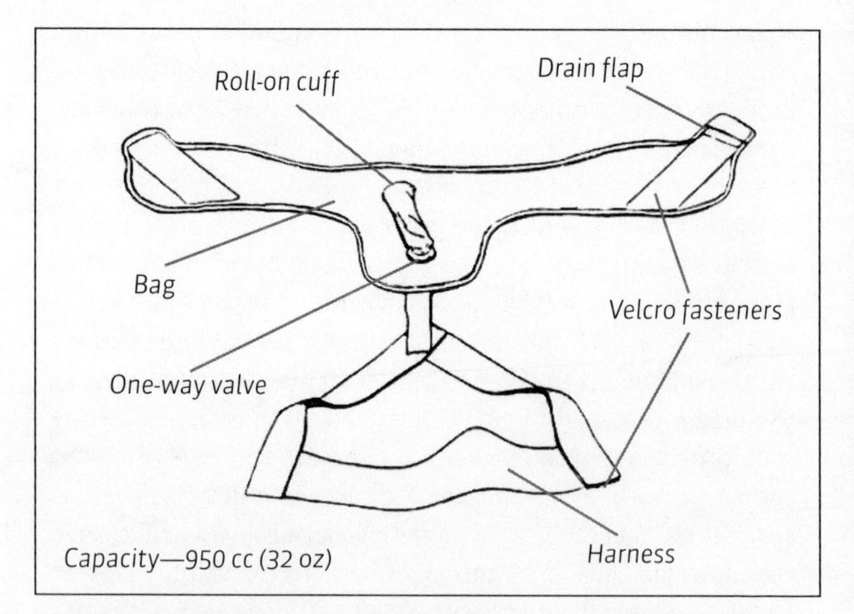

Spacesuit Waste Entrainment Temporary Cachement System (WETeCS). Ask for maximum capacity.

[3]

How to Choose Your First Ground Rover

So you're on Mars, safe and solvent, dressed up for action with a spacesuit you can count on. Yet you are still not ready. Before you can do anything interesting on the Red Planet, you're going to need wheels.

Mars is a very big place; its vast and spectacularly varied terrain covers a surface area equal to all the continents of Earth put together. And while at first you will probably only find yourself engaged in operations within the region surrounding a single settlement, the distances you will need to be able to traverse—between the spaceport, the habitation, agricultural and industrial domes, the nuclear-power station, the geothermal and water wells, the mining sites, the mineralogy and microbiology scientific zones, and numerous other widely scattered nodal points of potentially lucrative activity—are enormous. If you wish to engage in field exploration, prospecting, or claim-staking—which are still some of the best ways to make really serious moolah (through primarily legal means) on Mars—you will have to venture out much farther still. You simply can't do this sort of thing on foot. And if you aim to benefit the community through extralegal commercial activities (which, needless to say, for the record, I don't officially advocate), the wisdom of having a personal means to travel far and fast speaks for itself.

So the bottom line is this: If you are going to get anywhere on Mars, you will need to have your own motorized transport, and the sooner you get it, the better. That said, what type of ground rover should you choose?

Best and Worst Brands

Well, before I answer that question, I first need to warn you what type *not* to choose. So here's your warning: Don't choose a Boeing, Lockheed, Astrium, Arianespace, or any other brand that was developed with space-program funding and tested for Mars use on the moon. That's right, if somebody shows you a ground rover, and it bears the proud NASA stamp MOON-TESTED FOR RISK RETIREMENT, don't buy it. Don't even *think* about buying it, no matter what kind of bargain he offers.

Moon-tested vehicles are junk. While way back when NASA sold their lunar base program with the "rationale" (i.e., excuse) that it could be used as a testing ground for Mars exploration technology, in reality Earth's moon is the worst place (other than orbital free fall) for testing out a Mars car. The reason is that the moon only has about 40 percent the gravity of Mars, so during testing on the moon, such vehicles do not carry the same loads and are not subjected to anything like the kind of beating they face when they are used on Mars. And since the costs of space launch dictate that anything used on the Moon be built as lightly as possible to serve its function there, the structural elements of moon rovers are pitifully underdesigned for Mars use. Yes, you can drive them around the showroom just fine, but take them out on the rocky planitia and subject them to the repeated shocks of fast travel over rough terrain, and the car will literally start to bend out of shape in no time. I've never known one to last more than two months when subjected to any kind of serious use; most become undrivable after the first week.

No, instead of a rover allegedly pretested on the moon, you want one proved tried and true on *Earth.* So get yourself a Honda, a Kawasaki, a Shanghai-GM, or any other brand designed and built by people who don't know jack about space but do know how to build stuff that will hold up to extended rough handling on a planet whose gravity is nearly triple ours. True, they are much heavier than they need to be, and so they cost a lot to get here. But, hey, that was someone else's problem, which they no doubt solved by slipping NASA the bill. What's relevant for you is that the things are here now, and they last practically forever, so there are plenty of good used ones available at reasonable prices.

Pressurized or Not?

Having settled whose rovers you should buy, we turn to the question of technical specifications. There are two fundamental types of rovers: pressurized and unpressurized. Which should you choose?

Pressurized rovers carry an encased cabin, so you can ride in them in your shirtsleeves. They vary in size across the full range of enclosed vehicles on Earth, from minicars to land yachts, with the most typical ones being about the size and general form of a small-end terrestrial sport utility vehicle. The great advantage of a pressurized rover is that you can eat while you travel in one, or blow your nose, or get intimate with a traveling companion, and you have a good shelter to stay in should you get stuck out on the planitia overnight. The downside, however, is that big and heavy as they are, they are very difficult to get unstuck after you mistakenly drive into a dust or sand trap (which you will). So if you are traveling with someone, it is essential to make sure in advance that he or she is someone you want to be on intimate terms with, because you will be for at least several days, as rescues in such situations are always a long time coming (and inevitably longer when your companion is no fun).

In contrast, unpressurized rovers have no cabin, so you need to travel in your spacesuit. They vary in size from open four-seat or two-seat carts (like the old Apollo lunar rovers), to single-person saddled vehicles comparable to terrestrial All Terrain Vehicles (ATVs) or motorcycles. While offering considerably less comfort than pressurized vehicles, unpressurized rovers are a much better choice if you are into exploration. The reason for this is that they bring you into a closer and more informal relationship with Mars.

If you are traveling in a pressurized rover and see something interesting, you have to make (or negotiate) a command decision to stop the car, have everybody inside put on spacesuits, pump down or vent the cabin, and then go outside to get your samples, after which you will inevitably track dust back into the rover, which will turn it into a coughing closet until you get back to the base and have a chance to spend three or four delightful hours scrubbing its interior clean. So, while possible, going EVA from a pressurized rover is not something you decide to do lightly, and therefore, in general, you just don't do it. In contrast, if you are out and about in an unpressurized rover and notice something worthy of further inquiry—a potentially lucrative mineral suspect, for example—you just zip on over and pick it up or whack off a piece with a rock hammer, toss it in your minitrunk, note the location in your MPS, and move on. Then, when you get back to the base, you can show your sample to one of the docs, who'll be delighted to certify it in exchange for a 10-percent share on the claim. That's all there is to it. It's simple, easy, and you'd be amazed to know how many newbie saps just like you have gotten very, very rich doing just this while out driving on some stupid errand for the boss.

So forget about pressurized rovers and the comfort they offer. If you wanted comfort, you would have stayed on Earth, hanging around the community hot tub all day with all the other welfare bums. You came to Mars to make your fortune, and your chances of accomplishing that will be next to nil if you spend your travel time hiding inside a pressurized rover.

So unpressurized is definitely the way to go. Among the unpressurized types, I recommend the solo varieties. Two-seaters cost more, and why should you pay to transport someone else? Let them buy their own wheels. Even more important, one-seaters are more nimble, giving you access to rougher terrain, and they are also lighter, so that if you do break through the regolith crust into a dust trap, or get caught among some rocks, they are much easier to pull or lift free.

So here's the traditional rule of thumb when choosing a Mars rover: If you can't lift it, don't buy it. This is not as tough a limit as it sounds; remember gravity here is only one-third that of Earth, and strictly speaking, you can probably get by if you can lift one end at a time, but you *must* be able to do at least that. This is why many of the gals on Mars, having limited upper-body strength, choose Harleys or other motorbikes over ATVs for their personal transportation. Admittedly, this has exposed them to greater risk of crack-up accidents, but on the upside, the greater speed they enjoy as bikers has afforded them unmatched opportunities to engage in lucrative enterprises frowned on by the bureaucracy. That is one of the reasons why so much of the colony's wealth has ended up in female hands. So gents, if you are looking for the main chance, you might want to take note, and muster the courage to do by choice what the ladies have done by necessity. Remember, on Mars (as everywhere else, but especially on Mars): Fortune favors the brave.

As to the issue of being caught out on the planitia at night in an unpressurized vehicle, this can be dealt with nicely by

Having fast transportation can open up business opportunities. Shown above, informal salvage enterprise recovering useful parts from abandoned NASA vehicle.

ATV-equipped explorers through the expedient of towing a lightweight pressurizable trailer when out doing long-distance work, but even bikers can manage by keeping an inflatable cocoon bag stored under the ruck seat. If you get stuck, just whip out the bag, lie down inside, zip it closed, open the vent valve on your suit pack, then *whoosh*, you've got a place to stay for the night. Note: Always be sure to take your ration box in with you *before* you seal the bag.

Cocoon bags come in solo and (for those interested in more than mere survival) two-, three-, four-, and even five-person models. I prefer the duet style, but hey, that's just me.

Propelling Your Rover

As to the choice of power source to drive the rover, you want to go chemical, not nuclear or solar. Yes, it's true that nuclear rovers get nearly unlimited mileage, and the engineers claim to have done their shielding calculations correctly, but take it from me, riding around with a radioisotope power source under the seat between your legs just isn't something you want to do. Solar power may have sufficed for the first robot rovers back in the twentieth century, but those things traveled at top speeds of less than 100 meters per *day*. You can do a lot better walking. And sure, you can increase the juice greatly by towing a power cart with a raft of photovoltaic panels on it (sticking out every which way to bump into rocks as you go along), but you have no power at night, and even in the daytime you can be left totally out of luck should dust storms cloud the sky (as they can for weeks on end).

So chemical power is clearly the way to go. There are a lot of options here, but the wisest course is to choose the same power system for your rover as you use for your suit. That way each can back up the other. This is why I like the methanol/liquid oxygen suit fuel-cell and life-support system so much. Methanol/oxygen fuel cells are an excellent way to power a rover, and if your suit uses the same technology, then your rover consumables become an *enormous* reserve for your personal life support. Even a bike with a 20 liter LOx tank carries enough oxygen to keep you alive for weeks, if necessary, while the fuel cells used by both systems also produce your drinking water. (Of course, to actually survive such an experience and emerge sane, you would also need to deal with other issues as well. Once again, this is why you should absolutely insist upon a max-cap diaper system for your suit, as well as always keep a hefty extra supply of permaseal dispose-all sacks within the storage pocket of your cocoon.)

[4]

How to Choose Your Homestead

The next question we need to deal with is how to choose a place to live. You could shack up in a bunkroom in one of the central dome dormitories, and no doubt you may have to do so for at least a few days or weeks after your arrival. But the dorms are not where you want to stay. Nobody but total losers lives there, so you won't make any good connections, and with a permanent dorm address in your ID listing file, you're likely to get branded as a loser yourself. Furthermore, there is no point to it. Staying in a dorm means pissing your money away on rent for lousy quarters, when you *could* be living in style in a hab of your own while making a bundle from its appreciation in Mars's rapidly rising real-estate market.

Of course, if you invest in the wrong hab, you will lose your shirt, your underwear, and possibly parts of your anatomy as well. So you must choose wisely. But don't worry; providing you listen closely to me, it's dead certain you'll do just fine.

So, let's start at the beginning. On Earth, they say the top three criteria for determining the value of real estate are location, location, and, you guessed it, location. The same prioritization of location is true on Mars, only it's the top five criteria.

Now, clearly, by far the most valuable real estate on the Red Planet is that within the pressurized domes. Unfortunately, by the same token, it's likely to be too expensive for you. So let's talk about the cheaper stuff.

Unclaimed, undeveloped, or unexplored land is still available on Mars in vast quantities, nearly for free. However, you can't set up your hab just anywhere. While you certainly don't need the luxury of a pressurized dome around your house, there is something that you absolutely do need, and that is electric power. If you've got juice, you can do anything. But without electricity, you cannot survive.

The folks at S&R's home equipment department will try to convince you that you can deal with this necessity by buying your own photovoltaic array, microwave power receiver, dynamic isotope power supply (DIPS), or nuclear reactor. While you may find these pitches appealing to your sense of self-reliance, you should not fall for any of them.

Powering a house on Mars with solar photovoltaics is a really bad idea. It's not even an economical way to produce electricity on Earth, where the solar flux is two and a half times greater than here. Worse, there are the dust storms to contend with, which can reduce the light energy received from the sky by a factor of ten for weeks or even months at a time. This means you have to overdesign your solar array ten times bigger if you want to be able to get an acceptable level of power throughout the year. So a properly sized home solar array has to be huge, and thus very expensive, if it is going to actually do the job. Furthermore, even without the dust storms, there is always some dust in the air, and this will constantly settle on your solar panels, reducing their efficiency. So if you go solar, you may well find yourself spending endless tedious hours on profitless EVAs scrubbing the dust off your gigantic home photovoltaic array with a hand brush. It's no fun, and everyone will regard you as a total fool for having reduced yourself to such a plight. Real Martians don't do panels.

The need to dust solar panels may expose you to ridicule.

As for signing up with a microwave power receiver, don't even think about it. The way these things work is that they have this constellation of power-relay satellites in Low Mars orbit, which receive power beamed up to them via microwave from the central nuclear power stations in New Plymouth and Tsandergrad, and then return it using precisely aimed phased-array antenna systems to the rectenna receivers of subscribers on the surface. This arrangement is a slightly more sane version of a bizarre idea originally offered by a twentieth-century fantasy writer named Gerard O'Neill, who wanted to power the Earth by putting enormous solar arrays in geosynchronous orbit and beam the power down from there. This concept was really crazy, because it is thousands of times more expensive to generate power in space than on a planet's surface, so in fact it makes much more sense to generate power on the planet and beam it *up* (i.e., buy where it is cheap, sell where it is dear) than the other way around. However, once you do have power aloft, it can in principle be distributed widely by downward beaming. Thus the idea of a power-relay satellite.

Of course, if transmitting, satellite relay, and ground receiving rectennas are to be kept to a reasonable size, the relay satellite cannot be in geostationary orbit (36,000 km high on Earth, 16,600 on Mars) but rather, because the required dimension of the rectenna goes in direct proportion to the transmitting distance, in low orbit just a few hundred kilometers up. But since such low-orbit satellites do not have the same orbital period as the planet's rotation, and thus do not maintain a constant position with respect to the surface, a constellation of at least twelve is needed in order to maintain constant customer coverage. Thus the design of the current orbital microwave power-relay system, which can, in principle, deliver electricity with ten-meter accuracy to users nearly anywhere on the surface of Mars.

It's true that this system does work well, most of the time. But there are two big problems with it. In the first place, they lose half the power transmitting it up, and half of the rest on the way down, so you end up paying four times the price for your power as someone who has a direct wired hookup to the reactor generator system itself. That may be OK for users at a temporary prospecting camp out in the wild, but it is not something you can accept for use in your home power supply, 24.6 hours per sol (that's our word for a day-night cycle), 669 sols per year.

As bad as that is, however, the second problem is even worse, and that has to do with the difference between the (generally true) claimed accuracy of the power beaming targeting system and its (occasionally true) significant inaccuracy. Mars is not a perfect sphere; it bulges out at Tharsis, and this, as well as other asymmetric features, creates gravitational anomalies that constantly alter the orbits of the relay satellites. Furthermore, the satellites' orbits and orientations are also affected by the miniscule, but nevertheless tangible, drag the satellites experience from the traces of Mars's upper ionosphere still present at orbital altitude—which can increase radically and unpredictably following the expansion of the planet's ionosphere during a major solar flare.

As a result, both orbital and attitude corrections are constantly needed to keep the satellite system in tune, and these adjustments are not always made in a sufficiently timely or accurate fashion. The errors involved may be tiny, but it does not take much to make a satellite 400 kilometers above the surface swerve its beam 50 meters from your power receiver rectenna to microwave-fry you in your hab instead. Then there's the issue of liability. What if instead of frying you, your power delivery beam fries one of your neighbors? Or what if they claim that it did, just enough to cause all their medical problems? Do you really want to let yourself get stuck in such a mess?

So forget about using beamed power in your home. Not only

that, if you move into a community, be sure that it has a covenant preventing anyone else from doing so either.

Dynamic isotope power systems (DIPS) are attractive in many ways. They are compact, reliable, and deliver power at predictable levels around the clock without regard to Martian weather or any other environmental factors. These features made them very handy for early Mars explorers, and they are still sometimes used by prospecting outfits for the same reasons. But because they derive their energy from the natural decay of radioactive materials, they are constantly decreasing in power output, and thus value. This makes buying a DIPS a bad investment for you. It's true that the NASA-certified units, which employ plutonium 238 for their heat source, have a half-life of 88 years, so their rate of depreciation is fairly slow. But as a result of their plutonium loadings, those things cost a fortune. The affordable Russian-made DIPSes all use either strontium 90 or cesium 135, or a combination of the two, for thermal power. This makes them cheap, since plenty of radioactive strontium and cesium is available from the waste of terrestrial nuclear power plants. But the stuff only has a half-life of 30 years, so it loses value rapidly. Furthermore, it is a strong emitter of gamma rays, and the home power installation and maintenance people will use this as an excuse to hit you for extra special handling fees.

From the technical point of view, there are no issues with nuclear reactors—that's why they are still the dominant form of central-base power on Mars. But even if you have the means to buy one of your own, I have to ask why you would want to, since for ten percent of the price you could buy yourself a very swank penthouse inside the New Plymouth dome, and have all the power you need from the town grid, plus a pressurized neighborhood, and enough krill left over to invest in a portfolio of high-yield bonds that would support you in style for life, and then some. But why even talk about it? Since you are reading this book, it is obvious that you are not in that class. So let's talk about the real alternatives. Where can an ordinary just-shipped stiff like you find affordable land with acceptably low-cost access to power?

Aside from the region inside the city reactor's radiological protection zone, which we will not discuss due to my publisher's attorneys' excessive concern over liability issues, the cheapest fully electrified land on Mars is to be found in the areas near the main settlement spaceports. This can be a good place to set up house, particularly if you are into free trade in off-manifest cargoes, as your proximity to the warehouse district will no doubt provide you with many opportunities to participate usefully in this important economic activity. I must say, however, that such a choice is not for everyone, as the frequently recurrent supersonic blasts from the rocket engines of the ships landing and lifting from the spaceport at all hours can be very jarring, especially to those who aren't already deaf.

Furthermore, while offering good access to ample business opportunities, the spaceport warehouse district is an otherwise disadvantageous place to put a hab *because no one will ever put a dome there*. This is a key point. Unless you are loaded with enough krill to fill the Mars Authority headquarters building, you can't afford to buy or put your hab inside an established pressurized dome. But if you are smart, you *can* put your hab in a place where someone is going to set up a dome later, and if that should happen, you'll become a krillionaire yourself. In fact, you don't even need them to actually build the dome; even a substantial increase in the assessed development odds for your site at the booking exchange can multiply its value many times over. Are you starting to get the picture? Anyone can get rich on Mars. You just need to use your brains.

So, where are the next domes going to go? Isn't it obvious? They are going to be put over a place where a tight group of habs is in position in the open right now. You could try to join such a semi-developed settlement, which of course will also have a power system already in place. But it's a gamble, since the site cost will be fairly steep, and with the development-booking odds already considered favorable, significant appreciation will only occur if and when a dome is really built. That could take some time.

In my view, rather than hope for the reality of development, it's safer to bet on the increase in its expectation. So instead of putting yourself in hock to get in with a community in its adolescent phase, I recommend that you muster the guts to join with some other courageous folk in the enterprise of giving birth to an infant one. Remember: Fortune favors the brave!

So, where should you found your future metropolis? In a place where power can be created.

Going to Where the Power Will Be

Nuclear fission or fusion systems can efficiently produce large-scale electric power. But they are high-technology, high-mass, off-planet imports, and thus much too expensive to be afforded by anyone but one of the primary settlements or organizations with substantial government funding. If you and your new friends want to strike out on your own, you need a way to create your own year-round reliable medium-scale power generator using low-tech, locally produced (and thus cheap) equipment. There is only one way to do that on Mars, and that is to make use of geothermal power.

The fact that Mars has a hot core has been obvious since the late twentieth century, when NASA's Mariner and Viking probes photographed very large-scale volcanic features on the planet. While some of these features were old, it quickly became apparent from lack of meteorite scarring that many of them dated to within the past 200 million years. Since the planet is over 4 billion years old, such a date amounts to the geological present, and implied the continued presence of molten magma deep within the planet. This suspicion was directly confirmed a few decades later, when the Mars Global Surveyor probe, orbiting between 1997 and 2007, took a set of before-and-after photographs revealing that *within that time span* there had been a transient outflow of liquid water from the side of a crater wall. Since the average surface temperature is too cold to melt ice, such an underground liquid reservoir could only exist if there was a geothermal heat source below.

This has indeed proven to be the case. The deeper you go, the hotter it gets. And fortunately, New Plymouth is located in a region in which there are many sites where 300°C water can be accessed within a two-kilometer drilling distance from the surface. This is not *entirely* coincidental. New Plymouth grew from the base that developed around the first landing site, and the *Beagle* was targeted there because several scientists on the site-selection committee demanded that the mission go to a place where the crew might be able to drill down to reach liquid water as part of the search for life. From what I hear, the vote was close, since the NASA bureaucrats wanted to send the mission to a safer landing site in the middle of nowhere instead. But, as luck would have it, for once, reason carried the day, and given the 200-meter depth limit on the early exploration rigs, the choice of a region that could allow *them* to access biologically interesting low-temperature water has placed economically interesting high-temperature fluid within practical reach of the kind of drilling gear *we* have today.

The key issue, however, is just *how* practical. There's a huge difference in cost in establishing a geothermal power system off of a two-kilometer well compared to a half-kilometer well, and the price to the user of the power generated necessarily follows in direct proportion. Now remember, this is Mars you are on, not Earth; you need to use electricity for a lot more things. On Earth, where air and water are practically free, and no home heating is needed in most places for at least half the year, people can get by with an average power usage of about a kilowatt. Here you are going to need at least triple that. So you don't want to get stuck paying two-kilometer-well electricity bills—and don't believe anyone who tells you that you must. Trust me; there are still plenty of good half-kilometer hot-rock sites out there.

Now, in principle, you could go exploring for them, and that is something you might want to consider for the future, as staking such claims is a good way to make a pile of money. But you need to establish your own hab as quickly as possible to take advantage of the rising market, so you need to get in with other people who are planning a settlement on top of a hot spot that they have already found. How can you do this? Simple: you have to prove your worth. Any group trying to create a new settlement will always be on the lookout for good recruits. If you've followed the instructions I've presented to you so far in this book, you've already gone a long way toward proving your usefulness by helping to bring in some necessary goods during your immigration. You just need to continue along the same lines. If the syndicate gals providing the purse for the development want you to make a little payback for admission by pitching in a bit here and there to assist operations with their off-manifest cargo business at the spaceport, do it. After all, it's only fair, as without their money, there could be no new settlements. So help them make more. It's the right thing to do, and in return, they'll see that you get the homestead site you need. That's the Martian way.

[5]

Choosing the Right Technologies for Your Hab

Now that you have obtained a good homestead, we need to discuss how to choose the best technological systems for your hab. These systems include structures, controls, and life support. Let's talk about them one at a time.

Structures

Martian habs come in three structural types; rigid, inflatable, and tensile.

Rigid habs have solid aluminum walls, domes, decks, airlocks, and all the rest, just like the good old *Beagle* used by Becky Sherman and her comrades on the first landing. In fact, if you visit the *Beagle*, which is still standing in the central square at New Plymouth, you can see how little modern rigid tuna-can habs have changed from the ancient variety. They were the best habitat structural technology NASA could come up with a century ago, and, whether from their innate perfection or lack of creative capacity on NASA's part during the period since, they are still the best. They are, however, expensive to transport across interplanetary space, and once they have been landed, nearly impossible to relocate across any significant distance. So the chances of you getting yourself a rigid hab are rather slim.

This brings us to the subject of the much cheaper and more portable inflatables. Made from polypropylene fabric reinforced by Kevlar, Spectra, or Nanectra webbing, these cheap, lightweight structures can be packed in a box and easily moved cross-country for on-site inflation. The outside atmospheric pressure on Mars of 8 millibar (mb) is practically negligible compared to the Earth sea-level pressure of 1,000 mb, or 14.7 psi. Thus the standard 340 mb (5 psi) internal air pressure used in Martian dwellings puffs them up so firmly that they seem as solid as rocks.

Unfortunately, however, they are not. It's only the gas pressure inside that makes them seem strong; should that be released they will collapse instantly. The folks at S&R who sell these things will try to dismiss such a possibility by showing you how the Spectra netting will limit the damage from a micrometeorite strike to a pin-hole that you can easily patch before substantial leaking occurs. But micrometeorites are not the only potential cause of catastrophic deflation (or even a major one, since most are burned up in the Martian atmosphere, anyway). A much bigger threat comes from overheating of the plastic fabric wall due to faulty electrical wiring installed internally, which can cause it to melt and then rip open big-time—or even worse, start a fire, which can initiate an explosive decompression that will tear apart such a hab within seconds. Beyond such accidents, which you might be able to prevent with due care and caution, there is the threat posed by the hab inflation system itself. All modern inflatable habs use computerized gas pressure regulation, which employs a set of sensors to determine internal pressure and then adds or vents gas automatically to keep the system within specified operating limits. This may sound like a good idea in principle, but it puts you at the mercy of a computer. What if there is a software error in the programming, or if some kid somewhere hacks into your system and changes the coding as a practical joke? (Such pranks are standard among our colony's ingenious youth.) If you are lucky, it will happen when you are not there, because having your house deflate around you is no fun at all. But even so, coming home to a collapsed hab after a long day's work out on the planitia can be quite stressful.

So if you do choose an inflatable hab, you need to disable the automatic pressure stabilization system. Completely. Disconnect the computerized actuators, and put a manual ball valve on the vent line, so that the *only* way for gas to be vented from the system is for you to do it yourself—by hand. Trust me, if you set things up any other way, you will regret it.

Coming home to a deflated hab.

This brings us to the third structural option, which is the tensile type. These systems are similar to the inflatable units, in that they are made of fabric that is stiffened into shape by gas pressure. However they also include an internal framework supporting the walls, decks, and dome, which is generally installed immediately after initial inflation. Should the gas pressure be lost, this framework will still hold the hab in shape, much in the manner of a tent. In former years, this framework had to come from Earth in the form of pricey advanced aluminum alloys, but nowadays, cheap Martian-made steel struts are generally used. Even so, the added framework feature will perforce make a tensile hab cost more than a purely inflatable unit of equivalent size. However in my view, the peace of mind gained from knowing when you go to sleep that your house will still be standing in the morning makes the extra cost more than worthwhile.

It is true that the addition of the internal framework also makes the tensile hab more difficult to take down and move to another location. But this is not necessarily a bad thing, especially if you and your partner are having relationship difficulties, as it limits the ability of your significant other to vanish with your house while you are out working should she or he suddenly decide to split.

Controls

Martian habs contain many critical systems, and all of these systems require controls. We've already discussed one example of this, the case of the automatic pressure-regulation system used by inflatable habs. However, there are many others responsible for the operation of everything from lighting, communications, heating, cooling, and power conditioning to water purification, air-dust filtration, sewage processing, and oxygen recycling. Because of the number of such engineering subsystems that need to be monitored and regulated, NASA engineers have developed control systems that allow all of them to be monitored and directed automatically by the hab's central computer. Such Domestic Data Determination and Direction Devices (D5s) are considered to be a significant convenience to homeowners, and per Mars Authority regulation 40123, parts G1.276 through G1.341, are mandatory standard equipment to be installed in every hab.

You see, they never learn.

I've researched this matter, and it is incredible, because as far back as the 1990s, NASA's astronauts were tearing their hair out in frustration because the bureaucrats managing the International Space Station program insisted on making the lights on the ISS controllable only through the station's computer, rather than through manual switches (which would have corresponded to standard practice in all countries on Earth, even the United States, at that time, since that country's current government-controlled municipal anti-greenhouse illumination curfew system was not yet in effect). The astronauts complained that they wanted to be able to turn the lights in the station on and off directly, without trusting a computer to do it for them, but the bureaucrats insisted that central computer control was the more modern way, offering greater convenience and so forth. Of course, the system proved to be a complete nightmare, with the silly computer randomly turning the station's lights on and off at all hours, interrupting both sleep and work to torment the astronauts according to its unpredictable whims. But, since the managers in charge did not have to endure the consequences of their decisions, the failure of the station's computerized lighting control was not taken as a lesson to abandon such approaches. Instead, new multibillion-dollar contracts were issued for second-generation central hab control processors, even more "advanced" than that on the ISS, for use in the lunar base. By the time the *Beagle* got to fly, they were on to the third generation, which would have completely destroyed the mission had the crew not been able to hot-wire the system. (If you look beneath the control panel of the *Beagle* exhibit at New Plymouth, you can still see the slashed and spliced wires around the D5, and the vengeful torch burns on the D5 itself.)

So now that you have your own hab and your own D5, you need to handle it the same way the crew of the *Beagle* handled theirs. Otherwise, you will be forced to endure living in a house whose lights, water, toilet, refrigerator, ventilation, cooking, cooling, and heating systems are not obedient to your will. Instead they will be switching on and off all the time in accord with the direction of your D5, either acting malevolently on its own, or under the inspiration left behind by your sadistic ex-spouse, or some teenage hacker in Tsandergrad, or some creep from the Mars Authority who just wants to help you by moderating your utility usage via remote control.

Freedom, as the ancient saying goes, isn't free. Disconnecting your D5 will cost you some time, as you will have to set the parameters or operate the controls of each of your hab's subsystems yourself. But if you want to be master of your own home, there is no other choice.

Life Support

All Martian habs are equipped with a life-support system, whose purpose is to make your life possible by providing you with oxygen and clean water, and removing your carbon dioxide and digestive wastes. There are two ways to do this, each with its own school of advocates. One method, known as biological or bioregenerative life support, uses living organisms such as plants or bacteria to generate oxygen and recycle human wastes. In some cases, the bioregenerative life-support system is taken a step further and is also used to provide some of your food supply as well. The other approach, known as physical-chemical life support, uses chemical reactors to split the oxygen out of carbon dioxide and pyrolyze or otherwise destroy digestive wastes. Since there are no plants to speak of involved, no food at all is produced.

I admit that the bioregenerative approach to life support has a great deal of esthetic charm, and I sincerely believe that someday the development of satisfactory hab systems of this type will be achieved. Earth, after all, is a bioregenerative life-support system, and it has sustained life for the past 3.5 billion years without difficulty. In a certain sense, taken as a whole, the New Plymouth and Tsandergrad networks of settlements and greenhouses are also bioregenerative life-support systems. That said, in my view, the many sad experiences of those who have chosen single-hab bioregenerative life-support systems overwhelmingly argue for using the physical-chemical approach instead.

The problem with biological life-support systems is that they are unpredictable. When you use one, you are betting your life on networks of microbes, plants, and animals, which have a very large capacity, and apparent inclination, to behave in ways other than those desired by the life-support system designers. This was discovered long ago on Earth in the simulated Mars station run in the western American desert by the privately funded Mars Society during the early twenty-first century. They tried composting toilets, greenhouse water purification, and other bioregenerative gimmicks, and all they got for their trouble was a stinky and probably unsanitary hab—this despite previous successful testing of the very same units elsewhere—because apparently it did not please the microbes involved to work very hard once they were relocated to Utah. In contrast, once they switched to using an incinerator toilet—essentially a rudimentary version of a physical-chemical system—the thing always worked whenever they had sufficient power to run it.

You would think that this experience might have made an impression on NASA, once they finally got their butts into gear, emulated the Mars Society, and established their own ground-simulation Mars station, but no. In fact, it was worse, because when they encountered the very same problems, instead of changing gears and abandoning the bioregenerative systems, they threw more money at them, since this was the only way to keep their internal biotechnological R&D (research and development) communities happy. As a result, over the past hundred years, billions of dollars have been spent on these things, and while they are now far more complex and advanced than their ancestors of a century ago, they still stink just as badly.

The unreliability of the biological life-support systems is due to their complexity, which is intrinsic to the system at every level, from the subcellular to the ecological, none of which is fully understood. This may be the modern age, complete with fusion reactors and interstellar probes, but it is still the case that no one really knows how a cell *works*, let alone a plant or a multiorganism ecology, and there are a million unknown factors that can affect their functions. The only reason why the Earth works as a bioregenerative system is that it is huge, and its vast size allows it to buffer itself against disaster. Sets of organisms whose local ecologies collapse are simply replaced by others moving in from the outside, and a new balance is obtained. But in a small system, this does not happen. This essential truth was also observed during the early space age, when the multiacre Biosphere II—a gigantic system relative to any biome that you might use to support your hab—was found to be far too small to stably support all the required biological cycles.

Physical-chemical systems, on the other hand, are fundamentally simple, as all they involve is running reactors that drive prespecified pyrolysis, oxidation, or reduction reactions under controlled conditions. This kind of operation is vastly more predictable than living systems. Provided you supply power and regulate the water to have the proper trace chemistry, an electrolysis unit will always produce oxygen from water, and it will do so at a precisely predictable rate. Illuminate a set of green plants with sunlight, and they may produce oxygen, or they may consume it, depending upon how they feel.

Another downside to bioregenerative systems is the amount of upkeep they entail. Do you really want to spend most of your free time working in a greenhouse? The salespeople for these systems sometimes advertise this aspect of the matter as a plus, with videos showing happy people puttering among the foliage, smelling the sweet perfume of the flowers. Well, I enjoy the natural aroma of nice flowers, and even keep a rack in my hab under artificial light for exactly that purpose. (I know that sounds hard to believe, but it's true.) But that is not at all how plants being fertilized by human feces and urine in a closed-loop bioregenerative life-support unit smell. Furthermore, it frequently happens that the pipes carrying your sewage into your wonderful little home biome become blocked, leaving you with the fun job of repeatedly cleaning horrible stinky slime from the system's filters and clogged plumbing lines.

As for the bioregenerative cleanup units' big talking point—that they can also be used to grow food—wait till you attempt to taste the harvest. I say *attempt*, because you will almost certainly throw up before the second bite. Look, if you like growing your own garden patch and can afford the gear to do it, that's fine. Get yourself a small greenhouse unit, and have a good time. If you are skillful and clever, and wisely choose to grow the kinds of crops that tend to be frowned on by the authorities who monitor the central agricultural domes, your little home greenhouse can become a source of profit and pleasure not only to you, but to your entire community. But *please* don't try to also use it as a sewage treatment plant.

(We will talk later about creative ways to use your sewage. For now, however, it is sufficient to note that you should not put it in your food.)

[6]

How to Save Money on Radiation Protection

We now come to the important subject of radiation. The natural radiation levels on Mars are about fifty times higher than those on Earth. This fact creates a great danger to new immigrants, as reacting to it with hysteria, many are induced to buy wildly overpriced radiation-protection systems for their homes. You don't want to be one of these chumps. So let's take a level-headed view of the matter, so you can see what you actually need, and not get taken for a ride by scam artists selling you things you don't.

Radiation doses are generally measured in units called Rems (but sometimes Sieverts; 100 Rems equals 1 Sievert) and can either be *prompt* or *extended*, the difference being whether they take place on a time scale shorter or longer than the weeks-to-months cellular reproduction and self-repair cycles of the human body. Prompt doses, like those received by hundreds of thousands of victims from the gamma-ray flash after the Iranian nuclear strike on Moscow, are very dangerous. If a group of people receive a prompt dose of 450 Rem, 50 percent fatalities can be expected, with the death rate increasing to 100 percent as dose levels rise to 1,000 Rem.

Extended doses, delivered at low rates over long periods of time, on the other hand, have no identifiable direct causal relationship to radiation sickness or death. Rather, on the basis of various studies projecting the observable consequences of medium-level doses fractionally downward to low-level doses, they are believed to statistically increase your risk of getting cancer at some point later in your life. This, of course, is something best avoided, since cancer treatments are ridiculously overpriced, and you don't want to put yourself in a position where a bunch of greed-ridden Mars Authority medics can soak you for a bundle. Still, looking at the matter as a business decision, you need to balance the amount of money you *might* lose that way against what you *definitely* will throw away if you allow yourself to be conned into overspending on unnecessary shielding.

So, that understood, what is the risk from extended doses? I don't think anyone really knows exactly, but according to the experts, such as they are, the rule of thumb is that it takes 60 Rem of extended dose to create a 1 percent risk of cancer within 30 years for a woman, while 80 Rem will create equivalent risk in a man. (Men are slightly less vulnerable, because they do not face the possibility of breast cancer.) If the length of time for evaluating effects is increased or decreased, the assessed risk rises or falls in proportion. So for example, the woman who receives 60 Rem or the man who gets 80 Rem will have a 0.5 percent chance of getting cancer within 15 years, or 2 percent within 60 years. Thus the older you are, the less it matters how much of a radiation dose you get, since the chances are you are going to die of something else first.

Solar Flares

Now, there are two kinds of radiation that represent health threats in interplanetary space: solar flares and cosmic rays. Solar flares come from the sun, in unpredictable fashion, with a typical incidence of around one big one every terrestrial year—although it is entirely possible to have two big ones inside of a few months and then nothing much for several years. Generally speaking, you can expect them more frequently during periods of solar max and less often during solar min, but in point of fact they can happen pretty much any time, and when they do, a big surge of radiation comes pouring out of the sun for several hours, with sufficient intensity to deliver a prompt dose of several thousand Rem to an unshielded space traveler. As noted earlier, such a dose is sufficient to kill, either immediately or after a fairly brief period of severe radiation sickness.

That's the bad news. The good news, however, is that the solar flare radiation is made up almost entirely of protons with energies on the order of a few million volts, and such particles can be effectively stopped by about 12 centimeters of water, or comparable mass (roughly 12 grams/square centimeter) in other forms, such as food, or things that water and food become as the flight proceeds. Interplanetary spacecraft, therefore, deal with this threat simply by using onboard provisions and wastes to create pantry/latrine storm shelters, with the ship's food, water, and sewage arranged about to offer sufficient shielding to weather a flare. Such shelters tend to be small, offering rather tight quarters during the few hours of a flare, but this is livable, and the experience can even be quite pleasant if the wastes are well sealed and you have taken the trouble to arrange in advance with your ship's purser to assign you to the same shelter as another person whose intimate acquaintance you might find desirable.

On Mars, however, we don't really need solar-flare shelters, because the atmosphere, thin as it is, provides about 21 grams per square centimeter of carbon dioxide mass shielding looking straight up toward the zenith, and an average of 65 grams per square centimeter if you average all the potential paths for radiation coming in slantwise from all directions. This is significantly more shielding than that offered by even the best onboard shelters used by first-class passengers on the fanciest spaceships, so it's certainly good enough for you.

Cosmic Rays

The other kind of radiation, however, is quite different. Cosmic rays don't come from the sun; in fact even now, some two centuries after their discovery, there still is no definite agreement as to where they do come from. However, come they do, as a steady stream of particles zipping into our solar system from interstellar space, with individual particle energies not of millions of volts, but *billions* of volts. That part is bad news, because it means you can't stop them with a dozen centimeters of water; instead several meters of water shielding would be needed to block it out. Since no ship can hope to carry that much mass, interplanetary passengers have to take the dose, which only amounts to about 30 Rem per year, or 15 Rem during a typical six-month Type 1 outbound transit.

On Mars, things are better, because the presence of the planet beneath you completely blocks out half the sky, and the average of 65 grams per square centimeter (equivalent to more than half a meter of water) of atmosphere around you cuts the remaining dose by almost half again. The bottom line is that your unshielded cosmic-ray dose on the surface of Mars will typically be something in the neighborhood of 10 Rem per year.

So what's the big deal? True, this dose rate is twice that prescribed as the maximum allowable for Earthling nuclear powerplant workers by the terrestrial health and safety bureaucrats, but who cares? You came to Mars to get away from those kinds of ninnies.

Let me walk you through the math. Let's say you are a 40-year-old woman. At 10 Rem per year, it will take you 60 *years* to accumulate enough radiation to give you a *10 percent* chance of getting cancer within the *next* 30 years, i.e., *by the time you are 130 years old, and almost certainly dead.* If you are a man, and thus a bit less vulnerable, you won't have to face the same modest level of risk until you are 150.

For anyone who can count, this elementary calculation should settle the matter. Unfortunately, terrestrial standards of mathematical literacy are so low that many new immigrants feel that they simply must get extra home radiation protection, and scam artists abound who are delighted to exploit such fears.

Their latest gimmick is lead-boride (PbB) shielding, which is supposed to be a great combination, since lead is good for stopping gamma rays while boron is excellent for soaking up neutrons. Thus PbB goes for top prices, a fact which, for the naïve, also serves to certify its value. Unfortunately, the most significant part of the gamma-ray dose comes from high-energy heavy atomic nuclei, and when these hit lead, they frequently create a burst of secondary light nuclei, neutrons, and gamma rays that collectively represent a dose threat several times greater than the original heavy nuclei did by themselves. So in fact, you would be better off living in a hab with no shielding at all than you are in one covered with very expensive 3-cm-thick PbB "radiation armor." Of course, if your PbB shielding were 30 cm thick, it would stop the secondaries, too, but who can afford that?

If you truly are radiation obsessed, a much better plan is just to cover the roof of your hab with a layer of sandbags, or at the outside, bags of borated water ice. The latter is probably the optimal solution, since the ice will stop the heavy nuclei with minimum secondary production, and boron works best as a neutron-absorber when it is used in combination with water—*not* lead. (This is because the hydrogen nuclei in water have the same mass as neutrons, and thus split the neutrons' energy in half every time they collide, as opposed to the heavy lead nuclei, which simply send the neutrons bouncing off like a ball hitting a wall. And it is only after the neutrons lose energy that they become vulnerable to absorption by the boron.) But because it can be made as thick as you please with negligible cost, at the end of the day, a sandbag roof covering can work almost as well. A good half-meter sand layer will cut your domestic dose rate down to 6 Rem per year, which should be low enough for anyone. If you insist on less, you could join with one of those groups who are installing their habs in the lava tubes. Then your radiation dose would be zero. But I ask you, is it really worth losing your home's glorious views of the rosy Martian dawns and sunsets just to shield yourself from a few lousy Rems? I don't think so.

[7]

How to Stay Alive in the Desert

More than a few immigrants have met their end in the Martian desert. So it is understandable that you, being a green-as-chlorella, fresh-off-the-gangplank Earthling bumpkin yourself, should view the prospect of working way out on the planitia with some trepidation. Yet work there you almost certainly must, if you are to avail yourself of the full array of highly lucrative opportunities that Mars has to offer. Therefore, you literally cannot afford to give in to fear, and fortunately, you don't have to. For in this chapter I will reveal to you the desert survival secrets known only to the most experienced Martian veterans. Master these, and you need feel no dread for anything the planitia has to offer.

There are four characteristic ways for newbies to get themselves killed in the desert. These include getting lost, losing heat, running out of water, and running out of oxygen. We address each of these in turn.

What to Do When You Are Lost

How is it possible to get lost on the Red Planet, where every spacesuit is equipped with a Mars Positioning System (MPS) receiver that provides precise latitude and longitude information? It's simple; MPS units sometimes fail. They can fail because of an electrical short, or because a cosmic ray hit induces a programming error on their chips, or because you hit them with something by accident and break them, or because they get old, or because a solar flare temporarily disables the MPS satellites themselves, or frankly, for no reason at all. Suffice it to say, they do sometimes fail, and if you do any serious amount of outdoor work, sooner or later yours will have a bad day too. If you are lucky, this will occur when you are on familiar ground where you don't need the MPS anyway. But what happens if an MPS failure catches you when you are truly out in the wild?

This can indeed be a dangerous situation. Deserts, like oceans, can have the character of looking pretty much the same wherever you are. They say that way back in the Second World War, before there was satellite-assisted navigation on Earth, whole regiments were lost in the North African desert and died of thirst. The same thing has happened many times on Mars, most famously to the much ballyhooed MPS-Magellan inaugural cross-country celebrity "cyber-navigation" demonstration tour. However, it doesn't have to happen to you.

The key is to be a real navigator, not an airhead "cyber-navigator" trusting your life to some piece of junk that the S&R sales force, in their lust for profit, decided to dispose of by selling to you. Being a real navigator means actually understanding the basics of the science of navigation, which should not be too hard, since they've been known for a thousand years. Think of yourself as someone confronted with the same problem as that faced a millennium ago by a perpetually drunk primitive Viking savage. If he could learn to handle it, so can you.

It's simple: every planet that rotates spins on an axis, which causes its sky to appear to turn around two poles, one in the north and the other in the south. On Earth, the north celestial pole practically coincides with the viewed position of Polaris, or the North Star, allowing it to serve as a useful mariner's guide for centuries. On Mars, we don't exactly have a North Star, but the north celestial pole is easy to find because it occurs at a point exactly halfway between Deneb and Alpha Cephei. Find that point, and you know which way is north. (If you can't find it, write your will, because both New Plymouth and Tsandergrad are in the northern hemisphere—if this point is not in your sky, you are a *long* way from home.) Not only that, by measuring its angular height above the horizon, you can determine your latitude. Now New Plymouth is at

Finding the Celestial Pole.

9 degrees, 24.31 minutes north, while Tsandergrad is at 21 degrees, 17.92 minutes north. So, if you are working out of New Plymouth (which I presume you are, because this book has not yet been translated into Russian) and you are lost, you just wait till nightfall, take out your sextant, and measure your latitude.

Did I forget to tell you to buy one? Well, no harm done. If you are reading these lines, you are obviously still alive. If you are too far north, go south, or vice versa, until you get to exactly the right latitude. Then all you have to do is head due east or due west, depending upon which side of New Plymouth you are on, to find your way home. If you don't know which side of the settlement you were working on (you really *should*), you can figure it out by using your watch, comparing the time you observe the sunrise to that in the daily schedule for New Plymouth (which I hope you have with you). If it rises early, you are to the east. If it rises late, you are to the west. If you can't figure out whether it is early or late, guess. Your chances of survival will still be fifty-fifty, which, under the circumstances, is certainly a lot more than you deserve.

What to Do If You Are Caught in the Cold

The above discussion of dealing with navigational failure presumes that you are still mobile when your system malfunction occurs. After all, if you can't move anyway, it hardly matters that you don't know where to go. But what happens if you are immobilized, in a place where you lack the shelter offered by a pressurized rover or a cocoon? In that case, you will soon have to deal with the daunting prospect of the frightfully cold Martian night.

I'm sure you've heard the stories about prospectors or other travelers whose vehicles broke down, and who then set out on foot, only to be found frozen into rock-hard ice sculptures where they finally fell, sat, or lay peacefully down in the –90°C darkness. Some of these frightful tales are actually true, as you can easily verify by looking at the statues themselves, the best of which are on permanent display at the New Plymouth Cemetery. However, while I don't want to sound unsympathetic, the fact of the matter is that most of these people got what was coming to them, because they were unbelievably stupid.

The Martian Dunce Corner.

I mean, really, folks, you have to realize that you are not going to get very far walking cross-country in a spacesuit at night. Doesn't it make a lot more sense to stay where you are and call for help? OK, you answer, but in some of the cases in question, the stiff's com or navigation systems also failed, so they couldn't just whistle up a rescue; weren't they right to attempt the night trek? After all, *some* people have pulled it off. Yes, and some of those people were actually right to try it, because they knew what they were doing, while others were wrong, but lucky. I will concede that it is not always easy to tell these two types apart (although it usually is). However it is obvious that all the stiffs were dead wrong, or they wouldn't be dead.

The only time it is worth trying to walk anywhere at night is if you know that you are very close to safety, you know where the haven is and how to get there, and you have good portable lights with you with *lots* of reserve power. You need portable lights to travel at night because, while Mars has two moons, they are too tiny to provide any significant illumination. So the nighttime planitia itself is literally pitch dark, and if you try to walk in it without your own lights you will fall, repeatedly, until you either break your faceplate and die of explosive asphyxiation/depressurization, or stop and sit down and freeze to death. You need plenty of reserve power for the lights, because unless you use the super-expensive NASA deep-space battery packs, which have radioisotope heating units (RHUs) installed internally, you will need to spend a great deal of fuel-cell or battery power just to keep your power supply itself warm. Otherwise, it will freeze and become nonfunctional, leaving you doomed in the dark.

If, however, your case is other than the ideal one identified above, your best bet by far is to stay with your vehicle. The smallest bike or ATV is still likely to have enough residual fuel and oxidizer in its tanks to let you plug in and power your spacesuit heaters on overdrive throughout the night. And even if its propellant tanks are empty, you can still use them to save yourself by ripping off their multi-layer insulation (MLI) and wrapping yourself up.

I have to say, MLI is amazing stuff. It's just a lot of thin double-aluminized Mylar layers with little spacers stuck in-between here and there, so that it seems very insubstantial. But in either a vacuum or a thin atmosphere like that of Mars, it does an incredible job of retaining heat. Mummy yourself in twenty or thirty layers, and your body heat alone will keep you toasty warm. You need, however, to make sure your fingers and toes (or, if you are in a skinsuit, other extremities), are adequately wrapped, or they won't be there in the morning. Also, it is very important to make sure that the exhaust from your spacesuit life-support system is released *outside* of the wrappings, because if not, the water vapor you exhale will get between the insulation layers, freezing your mummy wrap into a permafrost-MLI-mâché that can not only immobilize you completely, but utterly ruin the appearance of your statue when it is later found and retrieved for exhibit at the New Plymouth Cemetery.

Marooned Without Water

An alternative nightmare scenario would be to find oneself stranded out in the desert with sufficient shelter to survive the cold, but no water. To live, a human being needs to drink about a kilogram of water per day, with some water ingested at least every other day. On Earth, a wanderer marooned in the wilderness can frequently find lakes or streams offering all the aqua anyone could ever ask for, but there are no bodies of liquid water to be found anywhere on the surface of Mars. How then can you possibly hope to survive?

Actually, all it takes is a little ingenuity. If you have listened to my advice so far, you've chosen a methanol/oxygen fuel-cell system to power your spacesuit and rover. When these combine methanol with oxygen to make energy, they produce water and carbon dioxide as waste, with the water part comprising 45 percent of the total. So if you are wearing a suit power system with a 2 kg methanol fuel supply, this will combine with 3 kg of oxygen from your respiration reserves to produce enough water to support you for two days, after which it will still take you yet another three days to die of dehydration. If your rover tanks contain any remains of fuel and oxidizer, you can potentially use such an approach to last even longer. The water you obtain will include a significant quantity of carbon dioxide in solution, which is why NASA has banned systems that plumb fuel-cell wastewater directly back to the suit canteen. However, despite the claimed medical problem, it is a fact that in the twentieth century, many people chose to drink carbonated water as a matter of preference, and significant industries existed selling such products. (I am not making this up.) So you really can drink the stuff, and before you do any serious work out on the planitia, you should have your suit modified to make this possible. (They do a good job on this, as well as MPS beacon mods and other necessary suit upgrades, in the back room at Cheryl's Alterations, located just west of the NP spaceport. Tell them I sent you and you'll get 10 percent off.)

However if you could not afford a fuel-cell suit or rover rig (or were simply foolish and chose not to listen to my advice) you will not have such a backup water supply available. In that case, you will want to consider ways of obtaining water from Mars itself. Mars appears to be bone dry, but that is because it is so cold that all of the available near-surface water is hard frozen, resulting in an atmospheric water vapor pressure that is a couple of orders of magnitude lower than that prevailing in some of the driest deserts on Earth. Nevertheless, water is actually plentiful here. Randomly chosen Martian desert soil is typically 3 percent water by weight, and places can be found where there is actually as much as 60 percent water frozen into the dirt. This stuff can be gotten out by heating, with the two most common technologies employed being the "oven" and "greenhouse tent" systems.

Oven systems work in a pretty straightforward manner. You just shovel in the dirt, turn on the heat, and out comes the water. They require a lot of power, however, because you can't just melt regolith water, as it comes out of soil too saline to drink. Instead you need to turn it into vapor and then condense it in desalinated form, a process that takes eight times as much energy as simple melting. So the question is, where do you get the power? One answer is to carry a backup folding solar array in your rover. If the weather is clear, a 2-square-meter photovoltaic panel set can generate about 100 watts when the sun is well up, which is enough to evaporate 1 gram of water every 30 seconds, or 1 kg during 8 hours of prime daylight.

While workable, that's a lot of bulky, heavy, expensive, potentially balky, and highly stealable gear to carry around all the time, and especially inconvenient if you are a (smart) person who prefers to use a light ATV or motorbike for transport. So, many Martian explorers prefer to employ the alternative greenhouse-tent approach.

Greenhouse tents are just that; transparent tensile fabric structures that weigh almost nothing and can be quickly erected and placed on the ground wherever desired. What they accomplish is to warm the soil within a few centimeters of the surface a few degrees above zero centigrade, which is sufficient to cause much of its water content to outgas, given the low ambient Martian atmospheric pressure prevailing within the tent. If you have a typical 2-meter-diameter portable dome tent, it will cover an area of 3.14 square meters, which, at 3 percent soil water content by weight, will include more than 9 kg of water within 2 cm of the surface. The solar flux landing within the 3-square-meter tent carries a *thermal* power of over a kilowatt, so plenty of low-grade heat is available to make the water outgas from the soil, and plenty of it will, displacing thin Martian CO_2-dominated air within the tent with an atmosphere of water vapor. This vapor can then be gathered simply by sticking a white-painted copper plate under the tent, which, connected by a copper rod to another white-painted copper plate exposed to the cold outside the dome, acts as a cold finger to condense the tent's water-vapor atmosphere into ice. It is important that sufficient numbers of such condensation units of appropriate size be provided, as otherwise the soil's continuously outgassing water vapor can potentially blow the tent into the sky, or explode it, which is not the result you want. However, provided such matters are attended to, system operation is straightforward, and almost always failure free, as there are no moving parts, electrical systems, or software to go haywire. Of course, you still have to melt the ice off the condensers to use it, but this can be done using a small warmer-box requiring an order of magnitude less power than a regolith oven.

Because they are so much simpler and cheaper than photovoltaic ovens, greenhouse tents are frowned upon by the major aerospace contractors that control NASA. They are therefore lobbying the agency to ban them, and probably will succeed sooner or later. However, at this writing greenhouse tents are still legal. So I suggest you buy yours straightaway, as their price is sure to go up significantly after the decertification comes through.

Neither greenhouse tents or photovoltaic ovens will work, however, if good sunlight is unavailable, and this can be the case for weeks on end during dust-storm season. If you should find yourself stranded without water at such a time, you will need to fall back on your final resource, and get the water you need from *yourself*.

That's right. *You* are a source of water. Actually all people are, because water (along with carbon dioxide) is one of the two primary waste products resulting from respiration. But you are your best source, because if you find yourself stranded, only you will be there to help.

There are three output streams by which you, as a human animal, emit water: exhalation, urination, and defecation. We discuss the potential of each of these in turn.

Since your lungs are wet, so is your breath, with an included water vapor pressure of about 50 mb (5 percent of an Earth sea-level atmosphere) being fairly typical. So if you exhale at an average rate of 6 liters per minute, you will be releasing water vapor at a rate of 0.3 standard liters per minute, or 18 standard liters per hour, which is equivalent to about 360 grams of water per 24.7 hour sol. This water can be acquired fairly easily by attaching a condenser tube to your suit's breathing-system exhaust. However, it is only about a third of what you need to survive.

The additional amount required can be readily made up by utilizing your water from urination, which itself will generally exceed the kilogram-per-day minimum. In contrast to exhalation water, which is generally pure enough to drink untreated, urination water is not.

My publisher's legal department insisted that I include the previous and the following sentence for liability reasons. Please note that neither Random House, Inc., nor any of its subsidiaries, heirs, or assigns, advocates the drinking of untreated or improperly treated urine, under any circumstances, and will not be held responsible for the consequences should any reader of this book choose to indulge in such activity.

Therefore, in order to make use of your urination water, you will need a portable distillation unit to evaporate the water and separate it from the noxious components of the urine. Such "portastills" are essentially miniature versions of the water recycling system you have in your hab, and are available in a variety of sizes, styles, and colors either new from S&R or used at the prospector's exchange. (In this case, I recommend buying new, as there is frequently a reason why someone has decided to get rid of their old one.) Unfortunately, however, in scaling down from household to personal-suit-size purification systems, the duration of the treatment process applied to the urine is necessarily reduced, with the result that their water product, while drinkable, still tastes and smells a bit of urea. As this can be rather annoying, most people who use such systems employ added flavorings to counter the problem. I've gone through a number, including Traditional Tang, Earl Grey, and Cajun Spice, but found them all inadequate. So now I use Johnnie Walker, which seems to do the job.

Finally there is the water of defecation. This can represent quite a significant resource. The problem, however, is that it is thoroughly mixed with feces, which, in a fieldwork situation, will generally be found within the WetECs diaper unit inside your spacesuit. While, if you have a cocoon with you, you can take your spacesuit off and access the diaper, the process of mucking around with its contents inside the cocoon bag will quickly tend to make the cocoon itself uninhabitable. For this reason, NASA has developed at considerable cost field units for recycling feces water, which incorporate powerful heating elements into the diaper itself, allowing the vaporization of the feces water to be accomplished in situ, right under your butt. The emitted water vapor is then collected in a condenser bottle whose input line is affixed to the rear of the spacesuit much in the manner of a bobbed tail. Flavoring can then be added and the water recycled for personal use.

While this system is highly recommended by the Mars Authority, I have never known anyone to make use of it more than once.

Surviving Without Oxygen

We now come to what many new arrivals regard as the most fearsome scenario of all. What happens if you are caught stranded out on the planitia without oxygen? No doubt Earthlings view this predicament as particularly terrifying because it never happens on their home planet. However, while such feelings may be understandable, they are basically irrational, since oxygen is actually quite plentiful on Mars. You just need to know where to find it.

The most obvious place to get oxygen on Mars is from the atmosphere, which is 95 percent carbon dioxide. To get the oxygen out of the CO_2, all you need to do is react some hydrogen with it over a copper-on-alumina catalyst in a reverse-water-gas-shift (RWGS) reactor. This will yield water and carbon monoxide. The aqua you electrolyze to make your oxygen, as well as hydrogen, which you recycle back into your RWGS reactor to continue the process; while you just toss the CO back into the air as waste. (You can do that on Mars—we have no Environmental Prosecution Agency here.) Alternatively, if you find water, you can just electrolyze it to produce your oxygen directly.

These techniques are obvious and quite simple, but they do involve a problem in that, to produce the 1 kg/day of oxygen you need to live, the electrolyzer used by either of the above approaches will require an average round-the-clock power level of 180 watts. Unless you have a radioisotope generator with you, this in turn means that you would need a solar array capable of producing about 500 watts during prime daylight.

Well, if you are really scared of oxygen deprivation, you can go buy yourself a 10-square-meter photovoltaic panel set and make your own breathing gas that way—which, after all, is the same method by which it is done in the life-support system of your hab, or on an industrial scale at the central oxygen-generation plant at New Plymouth. But why waste good money on such a fancy (and heavy) approach when there is a much cheaper way to make do when you are out in the field? Really folks, safety is fine as far as it goes, but what's the point of keeping yourself alive if you have to spend so much to do so that you have nothing left over to use to have a good time?

So forget about making emergency oxygen by the book from the air or permafrost. There's an easier way that works just fine, and that is to use the regolith itself. Virgin Martian dirt is loaded with peroxides, and these can be made to break down and emit oxygen just by wetting them with water. This surprising fact was discovered by NASA's *Viking* lander probe way back in 1976. *Viking* was sent to Mars to look for life. One of its experiments involved wetting Mars dirt with water, to see what might grow. But the scientists got quite a shock when, instead of promoting a slow growth of native plants, the soil itself responded to its irrigation by immediately releasing a flood of oxygen gas into the test chamber.

Well, 1976 may be ancient history, but the trick still works. If you wet unprocessed Martian soil, you will get oxygen. So instead of a RWGS unit with a 10-square-meter solar array, what you need is a large plastic bag, a shovel, and a small roughing pump. To get oxygen, just shovel some dirt into your bag, and then wet it, using water obtained by the methods I explained to you earlier. In fact, highly saline water obtained simply by melting permafrost is also fine for this purpose. When the oxygen starts fizzing out, just turn on your pump to bring the gas up to your suit pressure, and inject directly into your helmet's auxiliary feed line. The stuff tends to smell a bit like fired gunpowder, but it's quite breathable. If the smell does bother you, you can deal with it by inserting a small activated-carbon filter into the gas feed line. When the fizzing stops, just dump the bag, reload with soil, wet with water, and continue. There's nothing to it.

[8]

How to Make Anything

If you live in New Plymouth or its environs, you can generally do best by focusing your efforts on your own racket and using the proceeds to buy everything you need. But if you wish to be part of the action in the new settlements or the opportunity-rich world of the Martian outback, then the old-time frontier skills still come in handy. Long-distance overland transport on Mars is very costly, and getting a direct-order supply delivery from Earth at a landing site other than the New Plymouth, Tsandergrad, or Taikojing spaceports is a project well beyond the will-to-spend of anyone but the Mars Authority. Furthermore, the Sisterhoods that run the informal trucking lines will be quick to jack their prices up even further if they sense that you are making use of their services out of need, rather than choice, while the MA bureaucrats who control the approval process for authorized shipping can be even more demanding. So if you want to engage in Red Planet pioneering without being forced to surrender all your hard-earned krill to such types, you are going to need to know how to make everything you need yourself. In this chapter I will show you how.

Fuel

Aside from oxygen, which your hab's life-support system will be able to manufacture for you from the Martian atmosphere using its RWGS system, the next most important consumable that every Martian pioneer needs to be able to make is fuel. There are a number of ways to go about this.

The cheapest fuel to make on Mars is carbon monoxide. As we have seen, CO is a by-product of the RWGS process that everyone uses to produce residential oxygen from atmospheric CO_2. It is thus widely available and frequently viewed as waste, but in fact it can be used as fuel in combination with oxygen in internal combustion engines, turbine generators, rocket engines, or even in special types of fuel cells designed for that purpose. It must be said, however, that it is a poor-quality fuel, with much lower energy content per unit weight than many alternatives, and is less storable and highly poisonous as well. For this reason, no one who is anyone on Mars ever uses CO fuel, at least not publicly, except possibly in emergencies, as to do so would be widely seen as indicative of either financial insolvency or general stinginess.

A much higher-quality fuel that is nevertheless quite easy to make is methane. You can make methane by reacting CO_2 with hydrogen in a Sabatier reactor. A Sabatier reactor is just like an RWGS reactor—both work fine at 400°C and a few bar pressure—only instead of using copper-on-alumina catalyst, you use either nickel- (to be cheap) or ruthenium-on-alumina (to be safe). With the catalyst switched, instead of producing carbon monoxide, the reactor will make methane. (Just in case you don't believe me, I've provided the equations that prove this in a table at the end of this chapter.)

Now the methanation reaction gives off heat energy, so if you are running a Sabatier reactor, you can use the heat it provides for other purposes, including driving water out of regolith, cooking dinner, or running an RWGS reactor, for that matter. In any case, what you get is methane (CH_4), which is *great* fuel, packing 5.5 times as much energy per unit weight as CO, so you can use it without any fear of financial embarrassment. You also get water, which you can split with your electrolyzer to give you useful oxygen, along with hydrogen, which you send back into the reactor to help keep it going.

The only fuel that beats methane in energy per unit weight is hydrogen, but to liquefy hydrogen you have to refrigerate it down to 20 Kelvin (twenty degrees above absolute zero). This is much harder to do than liquefying methane (115 K), oxygen (90 K), or even CO (80 K), and it takes a lot of power and very expensive fancy refrigeration equipment to pull off. Furthermore, if you do, all you get for your trouble is a fuel with one-fourteenth the density of water that you have to put in huge, overpriced high-tech tanks, whose large surfaces provide ample area for sufficient heat leakage to cause all your precious hydrogen to boil away before you ever get a chance to use it. That's why no one with any brains ever uses hydrogen fuel, and even NASA shies away from it for everything but deep-space or other very demanding missions.

Methane is particularly good as rocket fuel, which is why all the Mars-to-orbit ascent vehicles and Earth-return vehicles leaving the NP spaceport use it. If you ever get into true long-distance work, you can use it to propel your intercontinental ballistic hopper, but even as an ordinary prospector, you will need it to power your rocketplane recon drones. You can also use it in internal combustion or gas turbine engines to drive ground rovers, the latter option being particularly attractive if your line of business requires vehicles capable of higher speed than those of the Mars Authority.

However, for most ordinary ground-vehicle applications, methanol/oxygen fuel cells are preferred. This is so for a number of reasons, including the long engine life inherent in using a low-temperature power system, the ease of methanol handling and fuel-cell wastewater recovery, but most important, because of the powerful advantages, synergies, and additional safety backups resulting from using a power system common to spacesuits and ground rovers. So you'll need to know how to make methanol (CH_3OH) too. This can be readily done by filling a reactor with copper–on–zinc oxide pellets, heating it to 250°C, and feeding it with CO and hydrogen at 20 bar.

Mash two hydrogen molecules together with one CO molecule, and you get methanol. That's the idea, anyway. Unfortunately, however, the reaction doesn't get near completion on a single pass. So you'll need a recycle pump to send the unreacted gases back into the reactor again and again until they are all used up, but provided you do that, you'll get all the methanol you need to drive your little buggy wherever you like, no sweat.

Explosives

If you can make fuel and oxidizer, you can also make explosives, which you may need for mining, excavation, or as a potential adjunct to civil litigation, among other purposes. On Earth, they make stable solid explosives by mixing a fuel powder with a stable solid oxidizer, such as a nitrate or perchlorate. The chemistry for synthesizing the latter is somewhat involved, so on Mars we prefer to use pure oxygen, in either compressed gas or liquid form, and mix it directly with a fuel. Thus you can fill a 5,100 psi high-pressure bottle with 3,400 psi of oxygen and 1,700 psi of methane, and make an excellent bomb. Even better results can be obtained if you use liquid oxygen and liquid methane (mixed 2:1 by mole, or 4:1 by weight) because, by taking advantage of the higher fluid density offered by the liquid phase, you can pack three times as much explosive material in a bottle of a given size. Such methane/oxygen cryobombs offer nearly double the energy yield per unit weight as TNT, but they are not used on Earth because they are unstable and can be set off by the slightest shock or spark (as can the mixed compressed gas systems), and so the home-planet safety ninnies find them unacceptable.

Notice: Neither Random House, Inc., nor any of its subsidiaries, heirs, or assigns, advocates the construction or use of mixed methane/oxygen explosives, whether of gaseous or cryogenic liquid design, on any planet, under any circumstances, and will not be held responsible for the consequences should any reader of this book choose to indulge in such activity.

However this problem is easily dealt with by employing two-chamber bomb systems, in which each of the reacting ingredients is held in its own separate container until the moment for use arrives, after which a valve is opened or a partition is broken allowing mixture to occur. Only then is the system armed and dangerous. But it won't be for long, because once you mix it, you fire it. So really, what could be safer?

Plastics

According to a wise old Russian saying, "We're all naked under our clothes," and this truth no doubt explains why nearly everyone chooses to wear them. While on Earth there are still some anachronistic eccentrics who choose to garb themselves like barbarians with parts looted from the corpses of organisms, on Mars nearly all clothing is made using civilized synthetic techniques. Plastics are also necessary to make furniture, bedding, tableware, trash bags, storage containers, light equipment parts, and hundreds of other necessary or useful items. So unless you want to be either a sucker-spender forced to shell out big bucks for underwear, tires, bags, and chair imports, or a naked slob living on the floor of an unfurnished hab filled with loose garbage, you will need to know how to make plastics.

The most important plastics that every pioneer needs to be able to make are polyethylene and polypropylene. Polyethylene is adequate for most ordinary applications, including trash bags, plastic containers, and low-strength parts. For higher-quality equipment and good synthetic fabrics, polypropylene is the better way to go. Both, however, are made using the same basic chemistry.

You start by reacting methanol with itself to make dimethyl ether, chemical formula $(CH_3)_2O$, but we'll just call it DME for short.

The DME-making reaction produces just a little energy, but you can easily make it go using a 400°C, 1-bar pressure reactor filled with cheap gamma-alumina pellets. The DME you make is quite worthwhile in itself, as it makes excellent, clean-burning diesel fuel that, unlike the typical terrestrial petrochemical or biodiesel varieties, won't freeze at Martian temperatures. However, our interest here is making plastics, so we continue with the next step, which is to feed the DME into a second reactor, which is filled with the common zeolite catalyst ZSM-5. With temperatures between 400° and 450°C, and pressures between 1 and 2 bar, you can transform the DME into either ethylene (C_2H_4, at low pressure) or propylene (C_3H_6, at higher pressure). If you then heat either of these under still higher pressure, they will polymerize to form polyethylene or polypropylene, respectively.

Once you have your plastic, you can have it cast into any shape you like, or spun into threads to produce beautiful fabrics suitable for the manufacture of clothes in any style or color. (I recommend red-brown, as it matches the dust and thus reduces the need to wash frequently.) So why pay others for stuff you can easily make at home?

Bricks and Ceramics

While nearly all new settlements are begun using prefabricated hab modules, sooner or later you are going to want to expand your living space by building structures of your own. For this, you are going to need artificial building materials, and on Mars today, as in the unforested regions of Earth for the past 5,000 years, the simplest such material to produce is brick.

Making brick is easy. All you have to do is take finely ground soil or dust, wet it, put it in a mold under mild compression, and then bake it. The folks at S&R will try to sell you a fancy high-temperature electric kiln to do the baking, but it's really not necessary; you can produce perfectly acceptable bricks using a 300°C oven heated by cheap solar reflectors. If you want stronger bricks, just mix the soil before baking with some threads from old waste parachute material. This stuff is available at nominal cost from the spaceport syndicates, who have turned to everyone's benefit their ability to find expended lander parachutes faster than the Mars Authority recovery teams. If you really do need the extra strength you get from 900°C hot-fired bricks, the Sisterhoods can also help you with that, as they run a service placing bricking material under the heat-rejection system of the New Plymouth nuclear reactor. Their product is of excellent structural quality, if slightly radioactive.

(Note: The people who run this business have a constant need for new hires, and aggressively offer jobs to fresh immigrants. Do not accept.)

Descent-vehicle parachutes are typically released at altitude, and land well downwind of the settlement. If quickly collected and shredded, used parachute material can provide excellent brick reinforcement.

Gypsum, which is a mineralogical form of calcium sulfate, is plentiful on Mars. This is great, because all you have to do to get lime is to bake gypsum, and once you have lime, you can mix it with fine-ground dirt to make Portland cement, as good as the most expensive brands sold on Earth.

If you know that your structure is going to be used only in unheated applications, you can make very strong building blocks just by wetting soil and letting it freeze in a mold into hard permafrost. These permafrost blocks can then be sealed to each other using water "cement," which will freeze them together as solid as rocks. This is a very cheap and easy way to create buildings, but the hitch is that if the structure is ever heated, it will simply turn into mud and collapse. So if you ever consider buying a brick structure for your own use, you need to make sure that it is not one of these. Nevertheless, it is a very useful technique for making buildings for sale to others.

There is another important point that must be kept in mind when constructing a building on Mars out of brick, even if it is of the more durable baked variety, and that is this: bricks are only strong in compression. They have almost no tension strength. In other words, unlike steel, which can both bear heavy loads and resist forceful stretching, brick can only do the former, not the latter. This is a subtlety that is frequently overlooked by newbies, who make the error of thinking of brick as a "strong" construction material. It is, *on Earth,* because houses *there* don't have to contain a force that is trying to stretch them apart from the inside. But outside-the-dome houses on Mars do, in the form of internal air pressure, and if you try to put an atmosphere in an unreinforced brick house on Mars it will simply *explode*. This can be very annoying if you are the homeowner. So learn this lesson well; you must keep brick structures in compression. The cheapest way to do this is to just pile dirt on top of them, and use the soil weight pressing down on the brick walls from the outside to counteract the air pressure that will push out from the inside. To contain a standard Martian household atmosphere of 5 psi (340 mbar), a dirt layer 2.473 meters thick will do the trick. Of course, since not all soil is of precisely the same density, you may want to put on a little extra, just to be sure.

Clay minerals are everywhere on Mars, so making ceramic pottery is straightforward, using the same techniques employed widely on Earth since Neolithic times. Because their manufacture is so simple, clay pottery vessels were made and used in large numbers by the first settlers of New Plymouth. For daily use, they have since been largely supplanted by more durable containers made of plastic, metal, or glass. However, many still find making them a useful skill, as, with proper attention to detail, they can frequently be passed off as genuine antiquities to Mars Authority bureaucrats or visiting NASA officials for a significant profit. (For the record, I am totally opposed to the manufacture of cheap fakes for sale to tourists, as such items tend to undercut the market. For the sake of the community, it is essential that all those who choose to engage in the antique business keep scrupulously to the highest standards of verisimilitude, and price their products accordingly.)

Glass

The most common material on Mars is silicon dioxide, SiO_2. Comprising about 40 percent of ordinary Martian soil by weight, silicon dioxide is the basic constituent of glass, which thus can be made here using sand-melting techniques similar to those employed on Earth for thousands of years. Unfortunately for glassmakers, however, the second most common compound in our soil (about 17 percent) is iron oxide, Fe_2O_3, and it's in all the dust too. This poses a problem, because if you want to make clear glass, the sand used as feedstock must be nearly iron free, and such pure silica sand is hard to come by on Mars (but cheer up, there is no sand at all on the Earth's moon. The fools who settled *there* have nothing to work with but trash rock).

So, if you want to produce optical glass on Mars, you have a choice: you can either do some serious prospecting to find a natural quartz deposit to get extra-high-quality feedstock, or you need to remove the iron oxide from common soil. You can do the latter fairly cheaply by hitting the iron oxide with hot carbon monoxide "waste" from your RWGS reactor. The two will then react to produce metallic iron and carbon dioxide, after which you can remove the iron with a magnet. It's a tiresome process, but you do get to keep the iron for other uses, such as making steel, as I will explain shortly. (If you are not interested in making steel, you might consider just carting off some of the leftover iron-denuded material from the Mars Authority foundry at New Plymouth. The MA bureaucrats still haven't realized that this stuff would be good feedstock for their glassworks across town, so they leave the dump unguarded.)

Of course, not all glass needs to be clear to be useful. For example, having an iron-red tint in your product doesn't hurt if your goal is to make fiberglass, or other good glass-based structural materials. So, if you do get your hands on optical quality silica, you should be sure only to use it, or sell it, where it is truly needed, instead of wasting it on products that could do just as well with inferior grade feed.

Metals

The ability to manufacture metals is fundamental to any technological civilization, so it's definitely something that you will want to be able to do in your own settlement as well. Fortunately, in this respect, Mars is considerably richer than Earth, so you should have no problem providing yourself with all the metal you will ever need.

Steel

By far the most accessible industrial metal present on Mars is iron. The primary commercial ore of iron used on Earth is hematite (Fe_2O_3). This material is so common on Mars that it gives the Red Planet its color and thus, indirectly, its name. Earthlings have known how to reduce hematite to pure iron since the time of the Trojan War. There are at least two good ways to do it on Mars. The first, which I mentioned earlier, uses waste carbon monoxide produced by your RWGS reactor to rip the oxygen out of the hematite and make metallic iron and carbon dioxide. The other approach uses hydrogen to react with hematite to form iron and water.

Both of these reactions are nearly energy-neutral, so after heating the reactors to start-up conditions, neither requires much power to run. If you do choose the hydrogen technique, you will want to add a condenser to the exhaust to gather the product water. That way, you can just electrolyze the same water over and over again to keep making the hydrogen you need, so the only net input to the system is hematite. Carbon, manganese, phosphorus, and silicon—the four main alloying elements for steel—are very common on Mars, and specialty alloying elements such as chromium, nickel, and vanadium are present in respectable quantities as well. So once you make your iron, you can easily alloy it with appropriate quantities of these other elements to produce practically any type of carbon or stainless steel you want.

Carbon monoxide—how I love the stuff! The Earthling bureaucrats at the Mars Authority object to it because it is toxic, but as far as those types are concerned, everything is toxic. The fact of the matter is that the ready availability of carbon monoxide "waste" from RWGS reactors gives you the opportunity to perform kinds of low-temperature metal-casting techniques on Mars that are simply impossible to get away with on Earth. For example, you can take your RWGS exhaust carbon monoxide and combine it with iron at 110°C to produce iron carbonyl, $Fe(CO)_5$, which is a liquid at room temperature. Then you take the iron carbonyl, pour it into a mold, and heat it to about 200°C, which will cause it to decompose. When it does, very strong pure iron will be left in the mold, while the carbon monoxide will be released, allowing you to use it again. You can also deposit the iron in layers by decomposing carbonyl vapor, allowing you to make hollow objects of any complex shape you want.

You can also make similar carbonyls by combining carbon monoxide with nickel, chromium, osmium, iridium, ruthenium, rhenium, cobalt, or tungsten. Since each of these carbonyls decomposes under slightly different conditions, you can take a mixture of metal carbonyls and easily separate it into its pure components by successive decomposition, one metal at a time.

On Earth they have regulations that make this kind of advanced metallurgy almost impossible to practice, because both carbon monoxide and all the metal carbonyls are allegedly so "toxic." But really, who cares? Just don't breathe the stuff.

Aluminum

On Earth, after steel, the second most important metal for general use is aluminum. Aluminum is fairly common on Mars, too, comprising about 4 percent of the planet's surface material by weight. Unfortunately, as on Earth, aluminum is generally present here only in the form of its very tough oxide, alumina (Al_2O_3). In order to produce aluminum from alumina on Earth, they dissolve the alumina in molten cryolite at 1,000°C and then electrolyze it with carbon electrodes, which are used up in the process, leaving the cryolite unharmed. You can do the same thing here, making the carbon electrodes by pyrolyzing methane produced in your Sabatier reactor.

However, aside from its complexity, the main problem with using this method to produce aluminum is that it is a real energy hog (or very "endothermic," to put it in snooty scientific terms). It takes about 20 kilowatt-hours of electricity to produce a single kilogram of aluminum. That's why, on Earth, aluminum-production plants are located in areas where power is very cheap, such as the Pacific Northwest. On Mars, power is not cheap. At 20 kWh/kg, a 100 kW nuclear reactor can only produce about 123 kilos of aluminum per day.

So I say, why bother? Steel is a perfectly good material for building high-strength structures, and due to our lower gravity, steel on Mars weighs about the same as aluminum does on Earth. True, aluminum is desirable for special applications, such as electrical wiring or flight-system components, where its high electrical conductivity and/or light weight put a premium on its use. But for such instances, I recommend going for supplies to the spaceport syndicates, who offer a large assortment of excellent advanced aluminum alloys acquired through salvaging unnecessary components from Mars Authority landing vehicles left neglected on the tarmac overnight.

The efficient technique for making aluminum.

Silicon

In the modern age, silicon has emerged as perhaps the third most important metal after steel and aluminum, as it is central to the manufacture of all electronics. It is even more important on Mars, because by manufacturing silicon you can produce photovoltaic panels, thereby continually increasing your settlement power supply (providing you can find some poor sap willing to be stuck with the job of dusting off the panels). The feedstock for manufacturing silicon metal, silicon dioxide, SiO_2, comprises about 40 percent of the Martian crust by weight. In order to make silicon, you need to mix silicon dioxide with carbon and heat them in an electric furnace. The resulting "carbothermal reduction" reaction produces metallic silicon and carbon monoxide.

Once again, you can make the carbon you need by pyrolyzing some of the methane you make in your fuel-production reactor. The silicon-making reaction is highly endothermic, although not as bad as the aluminum-reduction reaction, and the overall energy burden involved in reducing silicon is not remotely comparable, because the quantities you need tend to be much less.

For some purposes, the silicon product from carbothermal reduction is good enough. For example, you can use it to make silicon carbide, a strong heat-resistant material (it's used in tiles to protect orbital reentry systems). However, any leftover hematite present in the reactor feedstock will also be reduced, and this will leave some iron impurities in the silicon product. To produce hyperpure silicon, good enough for computer chips and solar panels, you need to take another step, in which you bathe the impure silicon product in hot hydrogen gas, causing the silicon to turn into silane (SiH_4). At room temperature or above, silane is a gas, so you can easily separate it from hydrides of the other metals, all of which are solids. Then, if you want completely pure silicon, all you have to do is pipe the silane to another reactor where you decompose it under high temperatures, thereby producing pure silicon and releasing the hydrogen to make more silane. You can then dope the silicon with phosphorus or other selected impurities to produce exactly the kind of semiconductor device you need.

It is an interesting historical note that a century ago, some of the charlatans whom NASA employed to sell its lunar base program to the United States Congress did so by claiming that vast quantities of silicon could be produced on the moon and used to make photovoltaic panels there, whose power output would subsequently be beamed to Earth for consumption. This idea had many serious flaws, not the least of which was the fact that solar power could also be produced in Earth's deserts, for about one millionth the cost. But putting that aside, everyone should have known that the idea was lunacy because, while silicon dioxide is as common on the moon as anyone could ask for, the carbon and hydrogen necessary to turn it into silicon metal are both absent. While it's true that you can (and should) design the system to recycle these reagents, in reality such recycling is always imperfect, making large importation of carbon and hydrogen necessary. Then, if you put that fact together with the local absence of anything approaching sand-quality concentrated silica to use as raw material, it is obvious that the moon is practically the *worst* place to attempt to make solar photovoltaics.

Yet they are still trying.

Copper

There is copper on Mars. It is present in our soil at about the same concentrations that it is found in dirt on Earth. This is quite low, however, about 50 parts per million. So if you want to obtain useful quantities of copper, you don't extract it from soil. Instead, you go find places where nature has concentrated it in the form of copper ore. Commercially, the most important sources of copper ore on Earth are copper sulfides, and the same is true here. Sulfur is much more common on Mars than on Earth, and as a result, copper ore deposits can generally be found on Mars in the form of copper sulfide deposits formed at the base of lava flows. Once you find it, you can easily reduce copper ore by smelting or leaching, just as has been done on Earth since ancient times.

In fact, in general, the only way of getting your hands on enough of *any* geochemically rare element to be worth anything is by mining its concentrated high-grade mineral ore. But you will find such ores only where complex hydrologic and volcanic processes have happened that can concentrate these elements, and within our solar system, only Earth and Mars have experienced such processes. That's why Mars has ore and the moon doesn't. But unlike Earth, the best deposits on our planet have not been looted for the past 4,000 years by filthy primitives seeking shiny metals for their worthless trinkets. This gives lucky people like you the opportunity to be the first to discover concentrated ore here of nearly every delightfully rare metal necessary to those who wish to build a modern civilization—or, alternatively, to get very, very rich.

Technical Note (WARNING: High Science Content)
Equations for Making Anything

In order to facilitate your work in manufacturing fuels, plastics, explosives, metals, and semiconductors at home, I've assembled below chemical equations for most of the processes discussed in this chapter. These will tell you how much of each chemical you need to mix together to get the product that you want. The ΔH for an equation is its energy balance; if ΔH is negative, it means that the reaction is exothermic and gives off energy. If ΔH is positive, it means that the reaction is endothermic and thus requires energy to be driven forward. It can be seen, for example, that the RWGS reaction (1) is mildly endothermic, while the Sabatier (or methanation) reaction (2) is substantially exothermic. Since these two reactions can be done at the same temperature, you can use a Sabatier reactor as a heat source to *provide the energy to drive your RWGS system. Not a bad deal, huh? You get high-energy methane fuel and free power at the same time.*

Have fun!

Reaction	**Energy Balance**	**Product**
$CO_2 + H_2 => CO + H_2O$	$\Delta H = +9$ kcal/mole	(1) Carbon monoxide
$CO_2 + 4H_2 => CH_4 + 2H_2O$	$\Delta H = -40$ kcal/mole	(2) Methane
$CO + 2H_2 => CH_3OH$	$\Delta H = -23$ kcal/mole	(3) Methanol
$2CH_3OH => (CH_3)_2O + H_2O$	$\Delta H = -12$ kcal/mole	(4) Dimethyl ether
$(CH_3)_2O => C_2H_4 + H_2O$	$\Delta H = -10$ kcal/mole	(5) Ethylene
$3/2(CH_3)_2O => C_3H_6 + 3/2H_2O$	$\Delta H = -28$ kcal/mole	(6) Propylene
$Fe_2O_3 + 3CO => 2Fe + 3CO_2$	$\Delta H = -2$ kcal/mole	(7) Iron
$Fe_2O_3 + 3H_2 => 2Fe + 3H_2O$	$\Delta H = -2$ kcal/mole	(8) Iron
$Al_2O_3 => 2Al + 3/2O_2$	$\Delta H = +399$ kcal/mole	(9) Aluminum
$SiO_2 + 2C => Si + 2CO$	$\Delta H = +165$ kcal/mole	(10) Silicon

[9]

How to Grow Food (That Is Actually Edible)

The Mars Authority offers food from a wide variety of crops grown at its Central Agricultural Dome (CAD) at New Plymouth, but it all tastes like shit. This is because, despite the evidence of the results, the intellects in charge insist upon using the CAD as a means of recycling human metabolic waste. Better stuff is available from the greenhouses of the outlying settlements, but unless such a community is very near yours, the transport costs involved in getting their surplus to you can be quite large. Furthermore, if you make yourself too dependent upon your neighbors for food, they will be sure to soak you for everything you're worth when the occasion arises. For this reason (and to have the opportunity to take others for a ride yourself) it is important for your settlement to have the capability of growing its own, fully edible food.

To accomplish this objective, the first thing you will need is your own greenhouse domes. While small racks of ornamental plants can be grown using artificial light, the amount of electric power needed to illuminate plants to support any kind of serious indoor crop production is simply off the chart. Consider the following: the amount of sunlight that illuminates a single 100 hectare (250 acre) farm on Earth would take 1,300 megawatts of electricity to produce, which is roughly the same amount of power as that needed to run a city of *one million* people. On Mars, the lighting levels are only 40 percent as great, and plants do just fine with that—but still, who has 500 megawatts to spare to light a farm? That's practically half the present power capacity of the planet. No, the only way to grow crops is to use natural sunlight, and that means greenhouses.

All Martian greenhouses are inflatables, employing polypropylene plastic with an antiultraviolet coating. This film is reinforced by internal Kevlar, Spectra, or Nanectra netting, giving the bulk material a yield strength in the 200,000 psi range. The key difference between domes is their pressure rating, with units of various sizes being offered in the 68 mb (1 psi), 170 mb (2.5 psi), 340 mb (5 psi), and 1,000 mb (14.7 psi) classes. The lower the pressure rating, the thinner the material can be, the lighter the structure, and the less the cost. In the popular 50-meter-diameter model, a 170 mb dome upper-hemisphere unit needs to use a 0.5-mm-thick covering with a total plastic mass of 2 tonnes (metric tons), while the 340 mb unit requires 4 tonnes of 1-mm-thick material, and the others need more or less in proportion to their pressure. Which is the best choice for you?

The 68 mb units are attractive because they are so light and cheap—being just 0.2 mm thick and 800 kg in mass for a 50-meter-diameter system. Plants require only about 0.7 psi (or 50 mb of atmospheric pressure) to live and grow, so the 30 mb nitrogen, 25 mb oxygen, 12 mb water vapor, and 1 mb of carbon dioxide atmosphere provided by these structures is more than sufficient to give the vegetables everything they need to be happy. However, while plants can live in 68 mb atmosphere, you can't, and if you choose such a low-pressure dome, you will have to do all your work inside wearing a spacesuit. The increased workload that this entails is such a waste of time that, despite their low cost, the 68 mb systems are ultimately a bad deal.

If you want to be able to work without a spacesuit, you need to raise the pressure to at least 170 mb, which is the reason why "lite-air" units are offered at this rating. However, unless you are really desperate for cash, it really makes the most sense to just take the plunge and go all the way to the "medium-air" domes, which are suitable for operation at the same 340 mb pressure as your hab module. That way you can set up tunnels allowing you and your fellow settlers to pass freely between your habs and the greenhouses without having to mess around with endless compression/decompression operations, which, I have to tell you, really get tedious after a while. Furthermore, the 340 mb units make better farms, because with Mars's gravity at one third that of Earth, they provide just the right atmospheric thickness to allow insects to fly as easily as they do on the home planet. So if you go with medium-air units, you can have honeybees do your pollination for you, whereas the suckers who buy the lite-air systems end up getting stuck with doing that work themselves. This job is so boring that I don't know of a single settlement where a lite-air dome has not proven to be a total dog on the resale market.

In contrast, medium-air domes are terrific investments, because in addition to their immediate utility as settlement food-production units, they represent potential first-class pressurized living space, which can be easily gentrified and resold at a fantastic profit when the frontier advances and it becomes time for you and your fellow pioneers to move on. Even before that, having 340 mb greenhouse domes readily accessible from the habs of your settlement increases the value of the hab modules themselves, since such greenhouses provide excellent places for people to meet, especially at night when the farm staffers are not around. This makes settlements so equipped much more fun to live in than others, and the improved local quality of life will be inevitably reflected in the potential sale price of your home.

As for the "full-air" 1,000 mb units, forget them. These things are a NASA invention, apparently created because, in order to generate big-time payoffs for politically connected real-estate speculators, the old agency put its Earthside center for human spaceflight in the middle of a previously worthless sea-level Texas swamp. So the poor souls they stuck in the place came to believe that its thick, viscous, foul atmosphere was the norm that all human beings everywhere simply had to have. And while they were unable to impose this madness upon their fellow Earthlings, most of whom lived free of NASA control in healthier higher-altitude locations, they took solace in trying to force their heavy, gagging air on those within their reach, which is to say first the Loonies, and then us. But aside from its grotesque choking quality, the thick air concept is totally impractical because everything built to that standard has to weigh and cost three times as much as it would otherwise need to. Furthermore, the heavy air requires a huge nitrogen component that has to be made up at considerable expense when it leaks, and which exposes everyone who uses it to severe risk from decompression bends should they ever need to suit up and go EVA in a hurry. So, despite the ravings of the thick-air fanatics at Bush Ibn Saud (formerly Johnson) Space Center, the present 340 mb standard was eventually accepted by everyone else. Not being ones to give up completely, however, they still offer their 1,000 mb structures as optional equipment to those foolish enough to buy them. Just say no.

Once you get your greenhouse dome, you will need to anchor it properly. A 340 mb 50-meter diameter dome experiences an upward force of some 7,000 tonnes trying to tear it loose from the Martian surface, or 44 tonnes per meter of circumference.

Thus, to anchor the skirt of the dome to a strip of ground three meters wide extending all the way around the dome's circumference, and assuming typical Martian dirt with a density four times that of water, then the skirt would have to be anchored about 10 meters deep in order to have enough mass weighing down on an anchor strip at the skirt's bottom to secure the dome. To root a dome in this way, you need to dig a trench 3 meters wide, 10 meters deep, and 157 meters in circumference, bury the skirt, and then refill the trench above the dome skirt's anchor strip. But that's the hard way, as it requires moving about 5,000 cubic meters of dirt. The easier way is to dig a narrow, shallow circular trench (say 1 meter wide by 3 meters deep—just 500 cubic meters of digging), lay the skirt in it, and then stake the skirt into the ground with long, deep-penetrating, barbed stakes. These dome stakes, which are sold at S&R and worth every penny if they are actually up to spec, have pipes in them through which you can send hot steam after you have driven them into the ground. Once underground, the steam mixes with the surrounding dirt and the two freeze together into an extremely strong ring of permafrost, which roots the stakes, and thus the dome, firmly in place.

It's a very simple process, and certain to work perfectly, provided you make sure that sufficient numbers of fault-free stakes are correctly employed, and that the permafrost ring is adequately wide and deep, properly mixed and proportioned, gap-free, and completely and thoroughly frozen in all places before any pressurized atmosphere is allowed within the dome.

And that's it, although depending upon how long you intend to own the dome before you sell it and move on, you may want to provide additional protection for its materials in the form of a UV-blocking acrylic tarp, or even a rigid Plexiglas exterior geodesic-dome shielding structure. The installation costs for the latter are significant, however, and so generally best left for the second owners to handle.

Growing Your Crops

So now that you have some greenhouse acreage, the next issue is how best to use it to grow crops. The first thing you will have to decide is what kind of atmosphere to employ. With a 340 mb dome, you can set the oxygen at 200 mb and the nitrogen at over 120 mb, the same as in your hab, but the open question is what level to set the dome CO_2. Hab module atmospheres generally contain CO_2 levels of about 0.5 mb, equal to what now prevails on Earth.

Historical Note

It is a historical fact that prior to the beneficial effects of the Industrial Revolution, the Earth's atmosphere only contained about 0.28 mb of CO_2, and relative to current yields, the growth rates of plants were greatly stunted as a result of this impoverishment of their primary nutrient. On Mars we have the opportunity to take this improvement much further, by enriching the atmosphere in our greenhouse farm domes with still greater amounts of CO_2. A good choice is 2 mb, as this greatly increases plant yields without making the air stuffy. (Two millibar of CO_2 is actually the level that prevailed on Earth 60 million years ago during the Eocene Era, before the evolution and proliferation of grasses caused the catastrophic impoverishment of the atmosphere, reducing CO_2 concentrations to near-starvation levels for all other plants and sending the planet into repeated ice ages and associated devastating mass extinctions.) Human industrial activity was well on its way to restoring the earlier benign environmental conditions during the early twenty-first century, and in fact would have done so by now without any extra cost or effort, *if not for a global panic that erupted over the possibility of "climate change." This caused a bunch of treaties to be signed that stopped the atmospheric refertilization process right in its tracks. I know it sounds unbelievable, but I assure you that* I am not making this up.)

In addition to carbon dioxide, the other primary nutrient that plants need to grow is water. The warming of its own surface soil will put some water vapor into circulation inside the greenhouse dome, but you will need to supply more. This can be obtained either from your settlement's own geothermal well, or, if you are stuck with a dry hot-rock power source, by scavenging the nearby planitia regolith using the handy field water-acquisition techniques I taught you about earlier.

Once you have sufficient amounts of water circulating in the dome, the moisture will react with the peroxides in the ground, eliminating this hazard from the environment and releasing an initial supply of oxygen as well.

Beyond CO_2 and water, however, plants require a host of trace elements, which they generally get from the soil. In this respect, you are lucky, because Martian soil is generally richer in mineral nutrients than most land on Earth. You can see this clearly from the data in the table below, which is taken from an official Mars Authority publication, but reasonably accurate nevertheless.

Examining the table, you can see that with respect to the large majority of plant soil nutrients, Martian soil has everything you need to grow with, and more. True, your typical Mars dirt is poor in potassium, but you can get all you want of that in high concentration by going to the salt beds that abound on the dry shores of our planet's former seas, lakes, and ponds.

The physical properties of Martian soil are also favorable for

Comparison of Plant Nutrients in Soils on Earth and Mars

Element	Terrestrial soil (average)	Martian soil (average)
Nitrogen	0.14 %	Varies
Phosphorus	0.06 %	0.30 %
Potassium	0.83 %	0.08 %
Calcium	1.37 %	4.10 %
Magnesium	0.50 %	3.60 %
Sulfur	0.07 %	2.90 %
Iron	3.80 %	15.00 %
Manganese	0.06 %	0.40 %
Zinc	50 ppm	72 ppm
Copper	30 ppm	40 ppm
Boron	10 ppm	Varies
Molybdenum	2 ppm	0.4 ppm

(Source: "Penal Relocation to Mars: The Humane Solution," Mars Authority Publication #45712-81654G)

plant growth, since it is generally loosely packed and porous, and well adapted mechanically to supporting plants. Most of it includes a fair amount of clays. This is fortunate, as these are highly effective at buffering and stabilizing the soil pH level in the slightly acidic range, and also ensure a large reserve of exchangeable nutrient mineral ions in the soil due to their high exchange capacity.

The real main question mark is nitrates, which can be plentiful or rare depending upon your location. If worse comes to worst, however, there is always the 3 percent nitrogen portion available in the atmosphere. If you need to, you can gather this, pump it up to high pressure, mix it with hydrogen, and then fix the mix into ammonia (chemical formula NH_3) just by using your Sabatier methanation reactor with this alternative gas feed. That is, after all, how they make artificial nitrate fertilizer on Earth. (They also make nitrates for bombs, shells, and other horrific weapons that way, but no one on our more civilized planet would ever do that, since we get a much better explosive yield using easily made methane/oxygen mixtures.)

As for the nitrates in human feces, no self-respecting Martian settler would ever use them for fertilizer. The issue is not merely a matter of taste, or even of health, although if improperly processed such material can be a very dangerous source of disease. Rather, it is a matter of profit and patriotic pride. Because the Earth's moon is almost completely bereft of nitrogen, hydrogen, and carbon, the lunar colonists are desperate for any source of these materials and are willing to pay top dollar for imported manure. In fact, you can get more for dung at George W. Bush City (aka Loonipolis) than you can for precious metals in London.

So, for as long as the Lunar Authority's moon base retains its bloated budget, it can serve the rest of us as a kind of wonderful planetary-scale pay toilet, with the cash flow directed toward the depositors, rather than the other way. And while the moon is much closer to Earth than Mars, the energy required to lift payloads from the massive Earth is more than four times what is needed to take off from here, and this gives us a decisive advantage in the interplanetary poop trade.

Of course, the Loonies would probably be happy to buy actual food instead, but why bother to send those useless fully subsidized loafers the valuable fruits and vegetables that you had to work hard to grow, when it is ever so much more satisfying to make them pay through the nose for your shit?

Lunar agriculture, by the way, is a really entertaining subject, because in addition to lacking the water, carbon, and nitrogen needed to form the plants, the moon also lacks usable sunlight. That is, while the solar flux there is equal to that of the Earth, it comes in

Putting the lunar base to good use.

two-week-on, two-week-off pulses, which is unacceptable to most plants, so they end up being forced to use artificial electric lighting to grow their crops, at phenomenal cost. Also, because the moon has no atmosphere to shield its surface against solar flares, the wards of the Lunar Authority can't make their greenhouses out of 1-mm-thick Spectra-reinforced plastic, as we can. (They lack the raw materials to make either Spectra or plastic in any case.) Instead they have to use glass greenhouses, at least 120 mm thick, to protect their plants from the flares. These very heavy and expensive structures may appear to be quite robust, but they always develop cracks from thermal shocks caused by the moon's huge repeated day-night temperature swings, until they inevitably shatter and explode in spectacular fashion.

Aren't you glad you chose a real planet?

Beyond Fruits and Vegetables

Because you were brought up on Earth, you've no doubt had to endure years of lectures from goody-goody vegetarian schoolpreachers urging you to give up eating meat, because a hectare of corn can feed far more hungry people than a hectare of cattle forage. These arguments are nonsense on Earth, because the starvation of the poor on that planet is not caused by a global food shortage but by the corrupt governments which pay the veggie schoolpreachers to go around making everyone else feel guilty. On our planet, however, in an environment where we can't simply *take* tillable land already lying around in continent-sized tracts, but must *make* it with domes and our own hard work, the vegetarian thesis appears to have some merit. Our agriculture has to be efficient, and including large warm-blooded herbivores in the food chain is, in fact, very inefficient. Most of the energy of the plants these creatures eat goes into maintaining their body temperature, and very little ever reaches you. On the other hand, you can't eat the larger part of most crop plants anyway. For example, in the case of corn, rice, or wheat, you don't eat the roots, stems, or leaves. Instead, you are stuck plowing most of your crop back into the soil with the self-consoling thought that you are keeping the ground fertile. But if that were your true objective, you'd plow the whole plant back—you're really just wasting energy. So if you want to be efficient, you need to find a way to use the not-directly-edible parts of the plants, and the obvious way is to bring in some livestock.

Now, nearly a century ago, some scientists at NASA did a study of this problem, and decided that goats would be the key to future animal husbandry in space. There was some logic behind this conclusion. Goats, after all, are of convenient size, omnivorous, fast breeding, and milkable. Be that as it may, it's rather obvious that the scientists in question had never lived on a farm, because they clearly didn't have a clue as to what goats can do. Nevertheless, the study became Mars Authority gospel, and we are stuck with the consequences today. So here's an important tip: Don't let any goat near the plastic wall of your greenhouse dome. He'll eat it.

Don't get me wrong. I am not antigoat. On the contrary, I love a good goat steak as much as any other Martian, and greatly enjoy watching how much fun the children get out of playing with the critters. Furthermore, goats have repeatedly proven to be of great value when released in New Plymouth to keep the Mars Authority security patrols busy while important business is being transacted by serious people. It is indeed a truly wonderful sight to watch a squad of MA cops trying to catch a clever goat as it leaps gracefully over 3-meter fences in the light Martian gravity. Nevertheless, as a means of turning otherwise inedible plant matter into high quality protein with maximum efficiency and minimum labor, they leave something to be desired.

Then there are chickens, which actually do make some sense, because in addition to delicious meat, they provide a constant source of good protein via their eggs. Unfortunately, however, they can be rather messy. This has proven to be an especially annoying problem in those domes that have opted for full Earth atmospheric pres-

While occasionally troublesome, goats can also be helpful.

sure, since in Mars's one-third gravity, such environments allow chickens to fly and thus spew waste from overhead.

Of course, if all you want is protein per se, mushrooms are the way to go. One good thing that NASA did manage to accomplish way back in the late twentieth century was to isolate species of mushrooms that will grow on the waste portions of plants and turn 70 percent of their material into edible protein that is as high in quality as soy (which is actually better than either goat or chicken, although solely from a nutritional standpoint). The fast-growing mushrooms need no light, just a dark, warm room, some waste corn stalks, and a little bit of oxygen. So you can set up your own private mushroom ranch in any closet or underground cellar, and most settlers do, with a typical ranch consisting of a large room in the front for the standard protein-producing species, with additional chambers in the back for growing higher-value recreational varieties.

However if eating mushrooms bores you, and chickens or goats are too much trouble, there are alternatives. Cold-blooded herbivores, such as tilapia fish, are quite efficient in transforming waste plant material into high-quality protein. Fish farms on our desert world? Why not? You don't need a very large tank to grow tilapia, they taste delicious, and they won't escape to eat your dome. Furthermore, when the time comes to move on, you can drain the tank, send the fecal residue to the moon, and make a bundle.

The Basics of Success

[10]

How to Get a Job That Pays Well and Doesn't Kill You

I have included the phrase "how to get a job" in the title of this chapter because the cynics who infest my publisher's marketing department demanded it, as they wished to exploit for profit the obsessive insecurity that nearly all Earthlings feel concerning this issue. But now that you have fallen for their ploy and shelled out your precious lucre for the book out of the naïve belief that it might help you "get a job," you might as well know that I am not going to waste either my time or yours with such a completely irrelevant discussion.

Well, before you run back to the store and demand a refund, there are two things you need to know: (a) My decision in this matter is fully justified, and (b) this book is being sold under nonrefundable terms. So, since there is nothing you can do about (b), allow me to console you by explaining (a).

There is no point writing a chapter about "how to get a job" on Mars because *the problem does not exist*. I know that this statement must sound unbelievable to you, coming as you do from a planet where your primary lifelong concern has been to find a way to convince some institution to grant you a livelihood in exchange for service, but that is not how things work here. On Earth, you no doubt spent years or even decades of your life enrolled in innumerable programs whose advertised purpose was to "prepare" you for various professions but whose real purpose was to *exclude* you from such work until you had "paid your dues." Then, armed with a sufficient number of certificates of submission so acquired, you might hope to go forth and prevail in your desperate effort to beg someone to allow you to validate the potential utility of your existence through the gracious allowance of permission to do something useful.

On Mars, in contrast, no one is interested in blocking you from exercising your talent. If there is something useful that you can do, there is no web of red tape to hold you back. No one is going to ask to see your certificates before they "let" you work. You don't need anyone's permission to do anything; you just need to be able to do it.

This startling difference between the two planets in the way things are done is not because we Martians are so much more intelligent, quick-witted, open-minded, fair, and practical than Earthlings—though of course, we are. It is simply a matter of economics. Earth has a labor surplus, while Mars has a labor shortage. Since the time of the First Landing, the dominant reality of social life on our planet has been that there has always been too much work to do and not enough people to do it. Nothing is more precious on Mars than human labor time. That is why *none* of the anti-innovation regulations that tie up everything on Earth are in effect here. That's right, *none;* whether you're talking about the restrictions on "work destroying" technologies like meta-robotics or hypercrops, "methods stabilization" requirements, or the employment qualification and reservation laws, no one here—not even the Mars Authority—has the slightest interest in their enforcement. There is simply too much to do.

The situation, I am told, is directly analogous to that prevailing in old America, back in the period of its open frontier. Because land out west was there for the taking for anyone willing to strike out on their own, there was a perennial labor shortage in the more settled areas. This drove many employers to offer pay rates that were unmatched on the world scale, and to try to maximize the productivity of their workforce through the encouragement of technological ingenuity. Others, however, met the challenge of a tight labor market by recruiting indentured help from Europe through the offer of paid overseas transport, or even by hiring raiders to seize and enslave involuntary workers from Africa.

It's the same way here. The issue is not one of "getting a job." Believe me, if you want to work, there is no shortage of people here who would be delighted to put you to work. On Earth you were unnecessary. Here you are wanted. That *can* be a good thing. The key, however, is that you must choose the *right* job.

First, let's talk about some of the wrong jobs. Under no circumstances should you, as a new immigrant, accept employment from either the Mars Authority or one of the Sisterhood syndicates.

If you go to work for the Mars Authority, you will be generally viewed as a potential stool pigeon, and never be accepted in the mainstream of Martian society. And while it is true that middle- and high-level MA personnel can do fairly well for themselves by accepting gratuities in exchange for noninterference with unregulated activities conducted in the public interest, it is very unlikely that you will be placed in a position where you can avail yourself of such opportunities. Rather, they will probably put you to work cleaning out blockages in the sewer pipes in New Plymouth, and make use of obscure clauses in the fine print of your employment contract to impose "disciplinary" measures that keep you trapped in such work forever.

If you are male, it goes without saying that you don't want to work for the Sisterhood, because you will be limited by code to the three lowest levels of their thirty-three-level organization. That means nothing but low-paid muscle work until such time as your accumulated knowledge of their ways leads them to feel it would be best that your employment were concluded by means of some sacrificial diversion operation directed in the name of "the Higher Good." If you are female, your prospects with them are not much better; as a new immigrant who doesn't know squat, they are likely to see you as being useful for just one thing. If you want to do that, you can make a lot more working on your own. (For further details on the latter subject, see Natasha Charity Reynold's excellent work, *A Lady's Life Among Prospectors: How I Became the World's Richest Woman and Most Notable Philanthropist*, Random House, New Plymouth, 2113.) Perhaps later, when you have developed experience, property, and connections that would allow you to be taken seriously and offered entrance to their organization at the twenty-sixth level or above, you might consider accepting a position. But until then, keep your distance.

Note to any syndicate members who might be reading this book:

Dear Sisters:

Please do not take any of the comments in the previous paragraph amiss. As you know, I have always been both a good customer and a friend, voluntarily providing useful information, paying my bills promptly, and delivering favors upon request. By no means do I wish to imply that I believe that any of your activities are wrong. After all, business is business, and you know your business. And certainly I, as both a businessman and a Martian patriot, appreciate the value of your many fine endeavors, and most particularly, your prompt retirements of those MA officials who

> *choose to request gratuities beyond the accepted rates. Your vigilance in this area has done wonders to curtail governmental abuse, and all of us in the Martian community applaud you for it. Yet the newbies reading this book are my customers, and I feel it is my duty to both you and them to steer the men away from a course that might lead them to become liabilities to you. As for the women, surely you can see it is in your interest that I advise them to hold off joining until Mars has toughened them up a bit. Think of it as free pre-enrollment screening and training. After all, you don't want your organization glutted by weaklings. So peace, sisters, and onward toward the Higher Good.*

Whew, that was close. I almost forgot to stick that in.

To continue, I also recommend that you decline any job offer from S&R or similar corporations. They are dead ends and strictly for losers. Furthermore, if you are one of those for whom S&R has paid interplanetary transport in exchange for a seven-year indenture, I advise you to scrimp and save and moonlight other jobs to put together the funds to buy your way free of your contract as soon as possible. You didn't come to Mars to be a sales clerk.

So, if government, crime, and corporate servitude are all out of bounds, what's left?

Good, old-fashioned, honest, hard work, that's what.

Making Money Honestly on Mars

While government, crime, and corporate dronery may exhaust the list of occupational paths available on Earth, this is by no means true on Mars. Ours is a planet with a new civilization that is growing by leaps and bounds, and there are any number of things you can do.

In the first place, to take the most obvious, there is construction work. New settlements are going up everywhere on Mars, and the demand for construction and well-drilling teams is terrific. As a result, construction crew recruiters are everywhere, fiercely competing with each other for available labor. If you are willing to do some hard work, you're hired. As a newbie who desperately needs to learn the ropes, you should seriously consider accepting such an offer. By joining a construction crew, you'll make some good friends with good people and get solid training in how to use all the basic outdoor gear, as well as a number of specialized tools, all in a work environment where help is readily available if, or rather when, you screw up. The pay is not too bad either, and you can supplement your income significantly by selling off spare equipment to other teams when the boss isn't looking.

In just a few months of such honest, hard work, you will accumulate a set of skills that will place you in high demand among contractors all over the planet, enabling you to charge top rates for

If you get a spaceport job, be sure you know the launch schedule.

your time. Alternatively, you will have gained both the experience and nest-egg capital needed to join with a few of your pals to start your own construction team. After that, your future is secure.

But once you've learned how to handle yourself outdoors, you can do better—because if you're looking for quick money on Mars, the real action is not in construction but in prospecting. I mean really, why waste your time drilling a well when you can make as much, or more, simply by finding one? In fact, why bother looking for water at all, when there is gold, platinum, and rhodium to be found? Because there is.

Mars, like the Earth, has had a complex geologic history, including all the volcanic, hydrological, and microbial processes necessary to create concentrated mineral ore. Mars, however, today offers much greater concentrations of readily available precious metal ores than is currently the case on the home planet—because the terrestrial ores have been heavily scavenged by humans for the past five thousand years. In contrast, our world has remained pristine, unravaged by such barbaric pilferers, and thus is generally believed to still possess large surface or near-surface deposits of metallic treasure, including silver, germanium, hafnium, lanthanum, cerium, rhenium, samarium, gallium, gadolinium, gold, palladium, iridium, rubidium, platinum, rhodium, europium, and a host of others. All of these possess sufficient sales value on the terrestrial market that they could potentially be lifted to orbit and transported back to Earth and sold for a substantial profit. While this has not actually been done yet, the fact that it clearly *could* defines Martian precious-metal ores as a resource of tremendous value. All you have to do to get fantastically rich is to go out, do a little good old-fashioned prospecting work, and stake a claim, which you can sell off to others to utilize. And the beauty of it is that since actual mining operations are still well in the future, it doesn't matter how much precious metal your claim really contains, or if it contains any at all. Thus your success is virtually guaranteed. This is in marked contrast to the business of geothermal well prospecting, which is much riskier, as such claims may be subjected to attempted exploitation in the relatively near future.

It should be noted that in selling precious-metal mining claims based upon optimistic interpretation of the field data, you are not putting your customer at risk: Assuming he possesses any skill at all at his business, he should be able to resell the property at a considerable profit regardless. Far from harming anyone, by transforming previously worthless geographic locations into marketable property, you are *creating wealth,* thereby enlarging the pie of well-being that sustains all of humanity. Having done the work to confer this blessing, it simply stands to reason that you should be the one who gets the wealth in question first. The great metropolitan stock exchanges of Earth, which have done so much good—enriching millions of people with trillions of dollars by trading in companies that never produce anything—all work in accord with exactly the same principle.

If, however, you don't like field geology, are finicky about generating claims that are not necessarily completely proven by every redundant nitpicking scientific test that might academically be desired, or are some kind of holy roller who feels the necessity to actually deliver "useful" goods in exchange for the money you make (such confusion is the result of a poor religious education; the Bible clearly says you can't serve Mammon and God at the same time, and business is about serving Mammon), there are still plenty of ways you can get in on the prospecting action. One of the best is to go into the business of supplying the prospectors with replacement gear and supplies. You can easily acquire plenty of such equipment at low cost, simply by scavenging the camps of the failures, or alternatively, by picking quitters clean while they are sobbing around the spaceport before they ship back to Earth. Either way, prospectors who know what they are doing will be happy to pay you for the stuff with some of their claims, which you can then mark up handsomely and sell yourself with a clear conscience.

The Martian precious-metals prospecting business is thus a magnificent win-win enterprise, open to participation by people of every skill and character. A truly wondrous gift to our fair planet from the bounty of Nature, it promises to provide excellent income opportunities for all concerned for some time to come.

[11]

How to Fly on Mars

As we have noted already, Mars is a very big place. And while you can get started in life here with the kind of regional mobility afforded by a good ground rover, to truly succeed on our planet you will want to be able to take advantage of opportunities open only to those with global reach. Given the long distances involved, and the mountains, canyons, or other impassible terrain that frequently stand between points of economic interest, the only way to effectively deal with this problem is to fly. In this chapter I will explain how you can do so, cheaply, safely, comfortably, and with the least possible hassle from the bureaucracy.

The three most popular means of air travel on Mars are balloons, airplanes, and rocket hoppers. Each of these options has its own unique advantages and disadvantages, so we will discuss each of them in turn.

Balloons

The atmosphere on Mars at low altitudes is about the same density as Earth's at 100,000 feet. So, since balloons have been flying there since the early twentieth century, it has long been known that it is possible to travel by balloon here as well. In fact, since our planet's CO_2-dominated atmosphere has a molecular weight of 44, compared to Earth's nitrogen/oxygen mixture's 29, it is actually easier to fly balloons here than it is in Earth's mid-stratosphere. On Mars, not only hydrogen and helium, but any number of low- and medium-molecular-weight gases, including methane, ammonia, water vapor, nitrogen, carbon monoxide, oxygen, and even methanol can be used to generate buoyancy. Furthermore, lacking Earth's complex combination of continents and oceans as it does, Mars's global atmospheric circulation is much simpler, more consistent, and more predictable than that of the old world, and has been well charted for nearly a century. While it is (fortunately) true that the wind velocities near the Martian surface are generally low, they pick up quite nicely at modest altitudes, with 100 kilometers per hour being quite typical a kilometer or so above the deck. If you are willing to sail with them, these wind currents can provide you with a set of aerial highways that can whisk you from New Plymouth to Tsandergrad in days, or halfway around the planet in less than a week.

Of course, since Mars's atmosphere is quite thin, the size of the balloon you will need to employ will be a bit larger than typically used by sport balloonists on Earth. The Martian atmosphere has a low-altitude density of about 16 grams per cubic meter, and, as per Archimedes's celebrated law, a balloon can float a mass equal to the mass of the fluid it displaces. So taking into account the density of your lift gas and the weight of the bag itself, a good first-cut rule of thumb is that you can get about 10 grams of effective lift per cubic meter of your balloon. So, if you figure that you will need 300 kg of net buoyancy to lift you, your spacesuit, a cocoon bag, and a lightweight motorbike (which you will want, since you can't really expect to come down *exactly* where you desire), then a balloon volume on the order of 30,000 cubic meters—equivalent to a sphere 40 meters in diameter—will be required. This may sound huge, but if made of standard 10-micron-thick plastic film, such a balloon bag would only have a mass of about 50 kg.

For lift gas, you have a number of options. Hydrogen is the most effective choice, just as it is on Earth. Furthermore, on Mars, unlike Earth, it presents no fire hazard. It can, however, be hard to come by, since to displace 440 kg of atmospheric carbon dioxide you will need to make 20 kg of hydrogen. This in turn would require electrolyzing 180 kg of water, and since it takes about 5.5 kWh of power to electrolyze 1 kg of water, your total power bill will come to 990 kWh in all.

Given the cost of electricity on Mars, this is a bit pricey. (It's about the same amount of power most people use to run their hab round the clock for two weeks.) But worse, outside of a major settlement, you might not be able to buy so much juice at any price. So unless you know that you can perform your entire round trip with a single inflation, using hydrogen as your lift gas could be a problem.

Fortunately, however, there is a cheap alternative, and that is to use water vapor as your float gas. With a molecular weight of 18, good old H_2O can provide plenty of lift in Mars's –44 molecular weight atmosphere. All you have to do is make sure that the interior of your balloon is kept at a temperature above about 5°C, so that the water will maintain a vapor pressure greater than the 8 mb Mars ambient atmosphere. This can easily be done during daylight hours, simply by using a balloon that is colored black to catch the solar heat. At night, however, such a balloon, if left unheated, will collapse and crash after all of its inflation vapor turns to frost. Thus, if you are flying such a system, it is prudent to land well before nightfall.

Another popular technique is to shun even water vapor, and just fly using solar-heated CO_2 as your buoyancy fluid. This requires a hotter balloon than black coloration alone can provide, but a shiny aluminum metalized coating can send a balloon to over 50°C in Martian sunlight, which is good enough to enable flight. (A gold metalized bag can get to 90°C in the daytime, and thus provide even more lift. But such balloons tend to disappear while they are parked on the ground during the night, so I don't advise using them.) Because hot CO_2 is not as effective a float gas as hydrogen or even water vapor, such solar-heated Montgolfier balloons need to be bigger than other designs to lift the same mass, but, on the positive side, their inflation gas is as free as air (because it *is* air). Furthermore, since they can acquire all the float gas they need during flight, when you fly one of these babies, you can vent gas at will, allowing you to maneuver up and down in altitude to catch the best air currents to take you where you really want to go.

Despite the ease and convenience of such systems, there are those lacking faith in the certainty that the wind will deliver them to their destination of choice. For these, as well as for those worrywarts who turn into nervous wrecks every night on a balloon trip over concerns that a Martian surface breeze might brush their gas bag's 10-micron-thick plastic film against some nearby rock, causing it to rip wide open, winged airplanes provide an alternative option.

Airplanes

Winged airplane flight is possible in the thin atmosphere of Mars, just as it is in the similar environment of the Earth's mid-stratosphere. And just as on Earth, there are two good design options for such vehicles to choose from: subsonic or supersonic.

Despite their lower speed (about 700 kilometers per hour on

Mars), subsonic aircraft offer the advantage that they can use long straight wings, which provide them with a higher lift-to-drag ratio than the small delta wings employed by supersonic vehicles. In addition, they can use propellers, instead of rocket engines, for propulsion. These two factors make them more efficient for long-range flight than supersonic craft, although clumsier and more difficult to manage in terms of their potential interactions with surrounding terrain when they land and take off. In fact, given the lack of runways in most places on Mars, in order to be useful, an aircraft must be capable not only of horizontal flight but of vertical takeoff and landing as well. In the case of subsonic birds, this can be done using tilt-rotor systems, while supersonic rocketplanes

Ancient NASA design concept for a subsonic Mars airplane. Apparently they didn't know that the propeller needs to spin.

can make use of ventral jets that allow them to take off and land in similar fashion to the ancient British Harrier.

Rocket Hoppers

In this day and age, it may be asked, Why bother with wings at all? If you are going to use rockets to fly, why not just punch through the stupid atmosphere and do your traveling through drag-free space? The answer is that for flights of short and intermediate range, using a winged system enables greater range for less fuel than is possible using a ballistic system. Wings also give you greater maneuverability, since they allow you to effect course changes without using any propellant at all. This can be a significant advantage if, upon arrival over a distant rendezvous, you discover a reception committee of a different nature than that you had been expecting. Still, it must be said that for true transglobal flight, there is nothing like a suborbital ballistic rocket hopper that can take you halfway around the planet in less than an hour.

Yes indeed, speed is good. It's paying for the propellant that's the rub. From a performance point of view, using the same methane-oxygen propellant that the ascent-to-orbit vehicles do is your best choice, but unless you can arrange with the local Sisterhood to have it liberated from a nearby spaceport depot, it will cost you a fortune. For this reason, many of those engaged in private ballistic hopper operations have elected to go nuclear-thermal for their propulsion.

A nuclear-thermal rocket (NTR) engine is a very simple thing—not to be confused in any way with the huge and highly complex electric-power-generation, heat-rejection, and ion-drive-propulsion system combination utilized by an interplanetary nukey spaceship. Essentially just a mid-twentieth-century variation on a flying steam kettle, an NTR works by using a solid nuclear-fission reactor to directly vaporize and heat a fluid, which is then ejected out a rocket nozzle to produce thrust. The beauty of these things is that since the propulsion energy comes from the reactor, virtually any fluid can be used for propellant. If you want to maximize the engine's exhaust velocity, low-molecular weight fluids, like hydrogen, are best, and this indeed was the original idea behind the concept. Using hydrogen propellant, a good NTR can get an exhaust velocity of 9 km/s, which is double the 4.5 km/s achievable by the highest performing (hydrogen/oxygen) chemical rocket engines (and two and a half times the 3.7 km/s of the more serviceable methane/oxygen units). Thus a hydrogen-propelled NTR can deliver the same impulse as its best chemical competitor using half the amount of propellant. This is a very good thing, for example, if you need to lift propellant from the surface of Earth to Low Earth orbit, and are paying a fortune in launch costs for every kilogram so delivered. But in principle, any propellant fluid will work, even good old Martian CO_2. Now, it is true that having a high molecular weight as it does, CO_2 really makes a pretty lousy NTR propellant—delivering, at best, an exhaust velocity of 2.3 km/s, which is laughable by the standards of snooty NASA chemical rocket engineers. But the point is that on Mars, CO_2 is available everywhere, for free.

So let the snoots laugh. They are paying through their noses for their high-performance propellant. If you can get your hands on an NTR hopper, you can fly anywhere, land, and then refill your propellant tanks just by running an onboard air compressor. This not only saves you a huge amount of money in propellant costs, but opens up innumerable otherwise unavailable business opportunities; as such a system allows you to hop around the planet at will, without the need to check in with the authorities at any of our planet's overregulated spaceports.

Of course, NTR engines can be expensive, as they require bomb-grade fissionable material (93-percent-enriched uranium 235 or its plutonium equivalent) for their construction. This, however, need not be an insuperable obstacle. On Earth, back during the Cold War, nuclear bomb-making industries were set up, which, once started, required perpetual orders to maintain profits and employment. Thus, once they had enough bombs to kill every person on the planet, they came up with the theory of overkill, to justify the manufacture of expanded arsenals sufficient in size to kill every person on the planet two, five, and then ten times over. Then, when this imperative became too surreal to stand public scrutiny even in such dimly lit fora as the times provided, they came up with the idea of disarmament programs, under which thousands of nuclear weapons could be dismantled so that others might be built to take their place. This disarmament process has been going on for close to 150 years now, and as a result of such noteworthy efforts in the cause of peace, nearly unlimited supplies of excellent bomb-grade fissionables are currently available, in a variety of shapes, styles, and isotopic compositions. (At one point, someone had the idea of diluting this high-grade stuff with natural uranium to turn it into safe fuel for commercial nuclear-power reactors. But fortunately that plan was scotched when its potential harm to international oil prices was taken into account.)

This situation might be a cause for some concern to those cleared to know about its existence, but fortunately the good men and women who monitor these stockpiles are paragons of diligence and wisdom, and rarely are willing to make any part of their trust available to terrorists or other individuals whose intemperate actions might endanger public acceptance of the program. However, when job security is not an issue, they are quite willing to be reasonable, and obviously the shipment of modest amounts of excess old fissile material to Mars cannot possibly entail any danger to anybody. So, provided they are treated with the respect and generosity clearly due to people who labor so hard in the public interest, their happy cooperation can be readily obtained.

A number of the spaceport Sisterhoods have networks on Earth that include all the appropriate contacts. They can arrange the entire fissile transfer deal for you, end-to-end, as well as the engine material refabrication job that will need to be done on Mars in exchange for a piece of the action on your NTR hopper business operations. Their usual price is 60 percent of your first five years' profit, decreasing by 10 percent per year after that. Pretty steep, I know. But what you get in return is a vehicle that gives you fast-transit unlimited global mobility, completely free of bureaucratic oversight. That's something truly worth paying for.

Freedom isn't free.

[12]

How to Invest Your Savings

Once you have secured profitable employment, and with it, some degree of wealth, the question before you will be what to do with your growing stash of cash. You could just reinvest your savings in construction, prospecting, or peddling, or whatever traditional line of business you found your way into and thus presumably understand. You won't starve if you do so, and if you are one of those timid souls who are willing to settle for comfortable mediocrity, I could understand why you might choose to take such a course. But I advise against it. Because if you wish to make a really serious fortune, the smart move for you is to cut loose, move on, and invest all of your spare funds as quickly as possible in Mars's potentially vast new growth industries.

You must not hold back. If you do, you will miss out. The enterprises that will shape the human future forever are taking off on Mars today. This is your chance to get in on the ground floor, and play a part in scoring the biggest score ever on the bottom line.

I do not exaggerate in the slightest. The new businesses will yield super profits, and between them, Mars will achieve a position of unmatched dominance in the three key areas of *energy, thought,* and *space* itself.

Let me explain each of these in turn, and then you can decide for yourself where you want to put your money.

In energy, we've got it made, because Earth is shifting to fusion reactors, and we've got an unbeatable advantage as a supplier of the stuff that makes them go.

Deuterium, the heavy isotope of hydrogen, occurs as only 166 out of every million hydrogen atoms on Earth but comprises 833 out of every million hydrogen atoms on Mars. In other words, we've got a five-to-one advantage in this wonderful material, which not only is the fuel for fusion reactors, but can also be used to make heavy water, and if you have that, you can make fission reactors work without the need for any uranium fuel enrichment. Thus poor little countries that have been denied the right to self-defense as a result of unfair restrictions on international trade in uranium-enrichment technology can achieve their fondest aspirations via the plutonium route if they can just get enough deuterium. This makes the stuff very valuable.

The problem, however, is that you have to electrolyze 30 tonnes of ordinary terrestrial "light" water to produce enough hydrogen to make one kilogram of deuterium, and unless you have a lot of very low-cost power to burn, which nobody on Earth does anymore, the process is prohibitively expensive. Even with cheap power, deuterium is very expensive; back in the late twentieth century, its market price on Earth was already about $10,000 per kilogram in then-current dollars, roughly fifty times as valuable as silver or 70 percent as valuable as gold. That was in a prefusion economy. Since fusion reactors have gone into widespread use Earthside, deuterium prices have gone up tenfold relative to other commodities.

But on Mars, we constantly have to use most of our power in water electrolysis to drive our various life support and chemical synthesis systems. This means we can accomplish deuterium production here for *zero* or marginal additional electrolysis cost.

So, for example, if we apply a deuterium/hydrogen separation stage to the hydrogen produced by the electrolysis prior to recirculating the hydrogen back into a settlement's RWGS reactors, then every 6 tonnes of Martian water we electrolyze will yield about one kilogram of deuterium as a by-product. Now, every Martian personally requires about 2 tonnes of water electrolyzed per terrestrial year for life support, and roughly double this is typically used to support materials-processing operations. Thus, a New Plymouth–sized settlement of 300,000 people will typically electrolyze about 1,800,000 tonnes of water per terrestrial year. Once we add our deuterium separators, this will result in the production of 300,000 kilograms of deuterium per terrestrial year, enough to produce 4 terawatts (tW) of electricity, roughly one-third of what the entire human race consumed at the end of the twentieth century (or about 2 percent of what Earth uses today). At current Earthside deuterium prices, this represents an annual export income potential equal in value to two million kilograms of gold, or 7 kilograms of gold-equivalent income for each of the 300,000 inhabitants of the settlement in question. At today's quoted terrestrial average rate of $8.23 per kWh for electricity, the sales value of the power generated on Earth as a result would total nearly $290 *trillion per year.*

Are you starting to get my drift here? We're talking about serious money.

Generating energy is going to big business for Mars, but generating *thought* could be even bigger.

Consider this: We're a frontier society. Just as the labor shortage prevalent in colonial and nineteenth-century America drove the "Yankee ingenuity" flood of inventions, so the combination of our own extreme labor shortage and our practical technological culture has already led us to produce one invention after another in energy production, automation and robotics, and biotechnology. And this is just the beginning. The anti-innovation movement may be a disaster for Earth, but it is going to be a bonanza for us, because no one here is willing to entertain such nonsense for a minute. Just imagine: They've banned hypercrop research on Earth because the crazies there were panicked by the threat that experimental megatomatoes might escape from the lab and overrun the planet! Well, look what's growing in the Ares Botanicals greenhouse in New Plymouth—megatomatoes—and not flaky alpha-test types, but real, viable, proven crops. Do you have any idea what the patent for that strain is going to be worth when we license it on Earth? I'd say at least $100 trillion—and that's just one crop. Ares Botanicals has *dozens* of equally valuable new strains under development, with several key varieties nearing full marketability even as we go to press. It's obvious that everyone who invests their savings in Ares Botanicals is going to make a fortune. Why shouldn't you be one of them?

Straight Talk

In the interest of full disclosure, I will state for the record that I am one of the founding partners of Ares Botanicals, and hold 1.6 million shares, or 11 percent of the stock of the company. That said, it should be clear that there is absolutely no truth whatsoever to the malicious claims of our competition that I have been using this book to boost the stock for personal profit. Far from it. Rather, as a favor to you, dear reader, I am offering inside information about AB's imminent prospects to allow you to share in my good fortune. You may ignore it if you like, but you do so at your own peril.

Botanicals, of course, are just the beginning. As a result of fear-driven reactionary technophobic regulation limiting the scope of scientific investigations that can be legally conducted on Earth, we now have the opportunity to make good use of our Martian ingenuity to take the lead in metarobotics, nanotechnology, antimatter, cryonics, autocloning, cryptoweaponry, cyberstimulation, Jovio-pharmaceuticals, and practically every other field of modern, cutting-edge, progressive research.

The inventions we create in our centers of unfettered inquiry will revolutionize and advance human living standards everywhere, as forcefully and wonderfully as nineteenth-century American ingenuity changed Europe and ultimately the rest of the world as well. But more to the point, the licenses for those inventions, registered in the patent office on Earth, promise to yield an unlimited fountain of wealth forever to all of those smart enough to get in on the action today. And starting with your investment in Ares Botanicals, you can be one of them.

Beyond cashing in on energy and thought, we Martians now have the opportunity to access nearly infinite riches through securing control of the ultimate resource, which is to say the practical entirety of the physical universe, starting with the asteroids.

The asteroid belt contains vast supplies of very high-grade metal ore in a low-gravity environment that should make it easy to export to Earth. For example, let's consider a single small run-of-the-mill asteroid just 1 km in diameter. Such a body typically has a mass of around 2 billion tonnes, of which 200 million tonnes is iron, 30 million tonnes high-quality nickel, 1.5 million tonnes the strategic metal cobalt, and 7,500 tonnes a mixture of platinum-group metals whose average value is twice that of gold. There has never been doubt about this—since the nineteenth century people have examined thousands of samples of asteroids in the form of meteorites. As a rule, meteoritic iron contains between 6 and 30 percent nickel, between 0.5 and 1 percent cobalt, and platinum-group metal concentrations at least ten times the best terrestrial ore. Furthermore, since the asteroids also contain a good deal of carbon and oxygen, all of these materials can be separated from the asteroid and from each other using variations of the carbon monoxide–based chemistry we use for refining metals every day on Mars.

There are about 10,000 asteroids known today, of which about 99 percent are in the "Main Belt," lying between Mars and Jupiter, with an average distance from the sun of about 2.7 Astronomical Units (AU). The Main Belt group includes all the known asteroids residing within the orbit of Jupiter with diameters greater than 10 km. The remaining 1 percent, all small, are the Near-Earth Objects, or NEOs. The 1 percent figure, however, greatly overstates the proportion of NEOs to Main Belters, because their relative closeness to Earth and the sun makes them much easier for the overfunded loafers who infest the ranks of the terrestrial astronomical community to see. Of the Near-Earth asteroids, about 90 percent orbit closer to Mars than to Earth. Of the Main Belters, there are probably at least a hundred small objects remaining to be found for every big one the somnolent Terran nightwatchers have bothered to find.

As should be clear, these asteroids collectively represent enormous economic potential. The Near-Earth Object group is of some interest for future use in the service of planetary blackmail, but for mining purposes, the real action is going to be in the Main Belt, where *millions* of 1-km-class (7,500-tonne-platinum class!) objects undoubtedly reside.

If the Main Belt is filled with mountains of such high-value material for cash export, why has it remained uncolonized? The answer is simple: There is nothing there to eat. While water and carbonaceous material can readily be found among the asteroids (making them as a group far richer than Earth's moon), it is not necessarily the case that such frozen volatiles can be found on those asteroids that are most rich in exportable metals. Quite the contrary, the valuable metal-rich type M asteroids are nearly volatile-free. Moreover, while many of the Main Belt asteroids contain all the carbon, hydrogen, and oxygen needed to support agriculture, nitrogen is generally rare. Moreover, sunlight in the Main Belt is too dim to support agriculture, which means that plants there have to be grown by artificially generated light, which is completely impractical as a method of producing food. Finally, while collectively the asteroids might someday possess a significant mining workforce, it is unlikely that any one asteroid will ever have a large enough population to develop the division of labor necessary for real industrial development.

So while the riches of the Main Belt have beckoned for over a century, the establishment of mining bases there has proven impossible because of the need to support asteroid prospectors and miners with supplies delivered at excessive cost from Earth. But now that we Martians are ready to get into the game, everything is about to change. Because, as explained in clear, compelling prose in the Ares Asteroidal business plan (the relevant section of which I have reproduced for your education as a technical note at the end of this chapter), cargo payloads can be delivered from Mars to the Main Belt for one-fiftieth of the launch mass, and thus the cost, that is required to ship them from Earth.

In other words, we here on Mars have an unbeatable positional edge for reaching the vast mineral wealth of the asteroid belt. And now that the Ares Asteroidal company has been launched, including as it does an unmatched and highly trustworthy managerial team with a proven track record of start-up business success (it goes without saying that I am involved), we will soon have all the necessary ships and supporting infrastructure in place to exploit that advantage.

But that's just the beginning. When the time comes to make use of the resources of the outer planets, the Kuiper Belt, the Oort

Cloud, and the stars, we will be in position to get a lock on that business too.

Can you think of a better investment? Of course not.

So don't miss out. Ares Asteroidal stock is the best deal since Peter Minuit bought Manhattan for $24 (about $62 billion in modern currency).

Put your money down now.

With an unbeatable positional advantage against its terrestrial competition for Main Belt mining, Ares Asteroidal offers potentially unlimited profits.

Technical Note (WARNING: High Science Content)
The Mathematics of Main Belt Asteroid Logistics

Ares Asteroidal will enjoy an overwhelming positional advantage over our would-be terrestrial competitors in conducting trade to the Main Belt asteroids. We have the edge because our

Mars-based rockets need to generate much smaller velocity changes (ΔVs) to reach the asteroid belt than are needed by those leaving Earth, and as a result, we'll need to use much less propellant. Just how much less can be seen in table 1, where we compare the mass ratio (the ratio of the mass of the spacecraft including its propellant, to that of the spacecraft with its propellant tanks empty) required of our spacecraft leaving Mars with that of our Earthling competition.

Table I: The AA Technical Edge: Minimized Transportation Requirements to Anywhere

	Earth (would-be competitors)		**Mars (Ares Asteroidal)**	
	ΔV (km/s)	Mass Ratio	ΔV (km/s)	Mass Ratio
Surface to Low Orbit	9.0	11.40	4.0	2.90
Surface to Escape	12.0	25.60	5.5	4.40
Low Orbit to Lunar Surface	6.0	5.10	5.4	4.30
Surface to Lunar Surface	15.0	57.60	9.4	12.50
Low Orbit to Ceres	9.6	13.40	4.9	3.80
Surface to Ceres	18.6	152.50	8.9	11.10
Ceres to Planet	4.8	3.70	2.7	2.10
NEP round-trip LO to Ceres	40.0	2.30	15.0	1.35
Chem to LO, NEP rt to Ceres	9/40	26.20	4/15	3.90

(Source: "Ares Asteroidal Private Offering Business Prospectus," release 12.7)

In table 1, we've chosen Ceres as a typical destination, as it is the largest asteroid and positioned right in the heart of the riches of the Belt. You'll notice, however, that we've also given

Earth's moon as a potential port of call. Despite the fact that it is much closer to Earth physically, we can see that from a propulsion point of view—which is what counts if you are counting the beans—it is much easier for us to reach for the moon than it is for the Earthers! That is, the required mass ratio is only 12.5 going from Mars to the moon, while it is 57.6 from Earth—a nearly fivefold advantage for us! And that is delivering to Earth's own moon! Going to Ceres, our mass-ratio advantage is overwhelming: 152.5 to 11.1!

We've based all the entries in table 1 except the last two upon a transportation system using methane/oxygen (CH_4/O_2) engines with an exhaust velocity of 3.7 km/s and ΔVs appropriate for the best trajectories employing high-thrust chemical propulsion systems. We've made this choice because methane/oxygen is the highest performing space-storable chemical propellant, and it can be manufactured easily on Earth, Mars, or a carbonaceous asteroid. Hydrogen/oxygen bipropellant, while offering a higher exhaust velocity (4.5 km/s), is not storable for long periods in space. Moreover, it is an unsuitable propellant combination for a cheap reusable space transportation system, because its costs exceed methane/oxygen propellant by more than an order of magnitude and its bulk makes it very difficult to transport to orbit in any quantity using reusable single-stage-to-orbit (SSTO) vehicles (thus ruling it out for any reliable, robust, low-cost, single-stage, readily reusable surface-to-orbit system of the kind that AA is developing). The last two entries in the table are based upon nuclear-electric propulsion (NEP) using argon propellant, available on either Earth or Mars, with an exhaust velocity of 50 km/s for in-space propulsion, with methane/oxygen used to reach low orbit from the planet's surface. How anyone could imagine using ultra-expensive nukey ships to conduct profitable transport to the asteroids is beyond comprehension, but we in-

clude it for completeness to show that even if the possibility of such a fantastic concept is conceded, we still have a huge mass-ratio edge.

So, looking at the numbers in table 1, you can see that if we stick with the realistic assumption of basing our business strictly on cheap, proven, chemical propulsion systems, then the mass ratio required to deliver dry mass to the asteroid belt from Earth is 14 times greater than from Mars. But this implies a much greater ratio of payload to takeoff mass from Mars to Ceres than from Earth, because all the extra propellant burned during an Earth-Ceres journey requires massive tankage and larger caliber engines, all of which requires still more propellant, and therefore more tankage, and on and on. In fact, looking at table 1, we can safely say that useful trade between Earth and Ceres (or any other body in the Main Asteroid Belt) using chemical propulsion is probably impossible, while for AA's ships operating from Mars, it will be easy.

So what about that fantasy nukey cargo ship scenario? Take a look at the table, and weep, oh Earthlings, weep. If nuclear-electric propulsion is introduced, Mars still has a sevenfold advantage in mass ratio over Earth as a port of departure for the Main Asteroid Belt, and that means they'll need much bigger and heavier (and more expensive) nuclear-electric engines to do the mission, which will cut their useful delivered payload to the bone. If this is taken into account, our top scientists' peer-reviewed calculations show, AA's ships launching from Mars will have a payload-to-takeoff weight ratio at least fifty times higher than that of any Earth-launched competition.

But wait, there's more: All this analysis so far assumes that the ships return from the asteroid belt without cargo. If the added burden of hauling enough propellant all the way out from Earth to return substantial amounts of asteroidal metal cargo without refueling at Mars is thrown into the mission requirement, then the prospects for our Earth-based competition become even more hopeless.

These numbers are irrefutable. Everything that needs to be sent to the asteroid belt that can be produced on Mars will be produced on Mars, not Earth, and AA is the company that will see that it gets there.

If you could not follow the preceding technical discussion, that's OK. All you need to know is that now is the time to invest. By seizing this historic opportunity to buy AA stock at its current low initial private offering price, you can become one of the founders of an enterprise that will not only deliver unmatched profits to all involved but also open the way to further rapid expansion of nearly every vital sector of our great young planet's growing economy.

Ares Asteroidal: Infinite Profits from Infinite Space.

[13]

How to Make Discoveries That Will Make You Famous

While taking care of one's material needs is necessary, it is not sufficient to create a full and well-balanced life. Money isn't everything. To be truly happy, it is also essential to be famous. As the desperate anxiety displayed by today's Terran masses shows to every sympathetic observer, the anonymous life is no fun at all. Thus, throughout the centuries, thinking men and women have sought inner affirmation through fame. Now that you have come to Mars, you have the opportunity to achieve success in this critical area of personal spiritual development. In this chapter, I will show you how to do so.

Let us start by asking a question: Who is the most famous person in the history of our planet? The answer, obviously, is Becky Sherman. It is her statue that stands on a pedestal in front of the old *Beagle* in Founders' Square in New Plymouth; it is her life that schoolchildren are made to study in excruciating detail; it is her pithy epigrams that are endlessly quoted by patriotic orators on the Twentieth of July and every other suitable occasion. Yet the question must be asked in all candor: *Why?* Sherman, after all, did not command the mission, Colonel Townsend did, and it was certainly he, along with Gwen Llewellyn, the flight mechanic, who were most responsible for the miraculous success of the First Landing. In her performance as a member of the crew, Sherman, at best, comes in with the middle of the pack. Yet despite that, and despite the fact

Becky Sherman stands in bronze in Founders' Square, justly immortalized for showing all subsequent Martians the way to fame.

that after the return to Earth, Townsend had a successful political career in which he became president of the United States (which was an important job at that time), it is Sherman whom everyone remembers today.

Why? Because it was she who first discovered life on Mars.

There you have it. Since the very beginning of human history on Mars, the one thing that people have really cared about has been the search for life. *If you want to become genuinely famous, that is the field in which you need to score.*

So, now that you understand the true importance of the subject, allow me to provide you with some background.

Mars, as we know, was once a warm and wet planet, possessing for its first billion years oceans and other liquid water environments entirely suitable for microbial life. By the late twentieth century, many scientists knew this (that is, those who weren't being blockheads), and they also knew, on the basis of ancient microbial fossils called stromatolites found in Australia, that there were bacteria on Earth within 200 million years of the time 3.8 billion years ago that liquid water first made its appearance on that planet. In other words, the surface of Mars was friendly to life for a period five times as long as it took life to appear on Earth after it was possible there. So, if the already-current theory was correct that life was not an exceptional or miraculous creation, but a natural development under suitable conditions via some orderly and predictable process of chemical complexification, then it followed that life should have appeared on Mars, too. And while Martian life might have gone extinct in the meantime, the mere finding of fossils on the surface would prove the theory, thereby demonstrating once again that the laws of science as they have been discovered to operate on Earth also hold true for the universe at large.

As above, so below, amen. I mean, really, how could it be otherwise? We're talking half a millennium after Copernicus here, folks.

Did those guys actually imagine that Earth might have *different* physical laws from the rest of Creation? How?

Well, as stupid as the whole discussion was, it got the Mars program started, so I guess we should be thankful. But really, the debate was all nonsensical, and in fact just plain phony, because the scientists involved also knew that there was, and had always been, natural transport of material both ways between Earth and Mars due to meteoric impacts. So anybody willing to think logically for three consecutive seconds about the matter must have known that there had to have been life on Mars 3 billion years ago, if from no other source than Earth.

So when they finally sent the *Beagle* out, the NASA brass knew they had the dice loaded, and were all set to astound the world with their *incredible discovery of a second genesis,* which they had scripted to occur on the Fourth of July, per White House instructions.

Well, being a reasonably competent field scientist, Becky Sherman had no trouble finding some stromatolite fossils by her third EVA, and she might have kept her mouth shut (although I doubt it) to announce the "discovery" in accordance with the political schedule, except that shortly afterward she found something that NASA was not expecting—life. These were cryptoendoliths, which is to say organisms that live hidden inside the surface layers of rocks, and why NASA did not anticipate them I don't know, as they had already been encountered on Earth in Antarctica and other extreme environments. But in fact, Sherman had made a really significant discovery, because her endoliths included not only the usual assortment of bacteria, but *free-living organisms simpler than bacteria.* Such "prebacteria," as she called them, do not exist on Earth, and in finding them, Sherman solved a fundamental scientific question that the terrestrial biological community had been trying studiously to ignore since the time when Louis Pasteur showed that spontaneous generation of microorganisms is impossible.

Bacteria are the root stock from which all other life on Earth has evolved. Yet, how can bacteria exist? They may be the simplest form of free-living life on Earth, but anyone who looks at their highly evolved structures, mobility systems, methods of replicating information, and other adaptations can see that they are obviously much too complex to be the first life to emerge from chemistry. Viruses are much simpler, but they are parasites *devolved* from bacteria; they require either bacteria or even more-complex cells evolved from bacteria to feed upon. So where did Earth's bacteria come from?

Sherman found out where. In discovering prebacteria on Mars, she didn't find the "second genesis" the NASA brass was looking for. She found the *first* genesis.

That's why her statue is in Founders' Square.

With Sherman's discovery, the mystery of the origin of life on Earth became clear. There are no predecessor organisms to bacteria on Earth for the same reason that there is no evidence of pre-Renaissance forms of Western civilization in North America—nothing more primitive was able to get there. Her examination of the prebacteria and their array of more developed descendants resolved other biological conundrums as well, most notably the absurdly idiosyncratic limits to the forms that biochemistry takes on Earth—all based as it is on just 20 amino acids from which are built RNA, DNA, and so forth. Why don't we see anything else? Well, on Mars we do. Earth life all has the same biochemistry because it is all descended from the one narrow subset of Mars life that made the trip across space first. Finally, by studying and classifying the prebacteria, Sherman was able to show several different stages in their progressive evolutionary history, thereby finally illuminating in a step-by-step manner the previously incomprehensible process whereby mere chemicals can develop into that miracle we call life.

Thus, simply by being in the right place at the right time and keeping her eyes open, a person whose second-rate talent and limited publication record would ordinarily have precluded her from ever getting a tenure-track faculty position at a respectable university was able to win herself eternal fame. The benefits from this included but were not limited to: (a) the Nobel Prize (a very prestigious award, which came with a large amount of cash), (b) numerous well-paid advertising product-endorsement contracts, (c) a nearly infinite number of invitations to high-priced speaking engagements, and (d) a series of marriages with wealthy movie stars, business tycoons, and European royalty, all of which ended with highly lucrative divorce agreements (the settlement on the last left her with the crown of Denmark).

So you can see how valuable an activity the quest for life on Mars can be. Clearly, this is something that you should get involved with too.

"But how can I do that?" you ask. "Surely all the great discoveries have already been made, either by Sherman or the few who followed immediately after?"

Not so. As you know, Mars is a planet with a surface area equal to all the continents of Earth put together, and in the century we've been here, we've done little more than scratch the surface. The vast majority of the planet remains unexplored. *Nobody* knows what is out there. So if you put in a little effort, the opportunity for you to go out into the unknown and make incredible finds is wide open.

Just to whet your appetite, here are some of the terrific fame-generating discoveries that you could make:

1. *Live nucleated cells of a type never before seen*
2. *Fossil traces of extinct multicellular organisms*
3. *Shells of extinct invertebrates*
4. *Fossilized bones of extinct vertebrates*

5. *Dinosaur footprints*
6. *Recently deposited wildlife stool droppings*
7. *Anomalous industrial chemical deposits in undeveloped locations*
8. *Tracks left by vehicles of unknown types*
9. *Surface blast marking patterns, not matching existing rockets*
10. *Rock inscriptions of unknown origin*
11. *Technological artifacts of unexplainable provenance*
12. *Ruins of structures clearly designed for nonhumans*
13. *Alien statues*
14. *Remains of pre-spaceflight humans, for example, medieval knights*
15. *The Holy Grail*
16. *The One True Cross*
17. *Angel skeletons*
18. *Books or scrolls made of gold foil, filled with writing that only you can decipher*

The above is just a partial list of sensational life-changing discoveries that you can make. It is by no means meant to be all-inclusive. Some of them require more work to make than others, and some, such as number 18, may be beyond your budget (although those who have tried it have generally found the rate of return to be more than ample to justify the investment).

The point, however, is that discoveries have to be *made*. Becky Sherman made hers her way; nowadays with the low-hanging fruit already taken, more advanced techniques need to be used. Fortunately, however, such techniques are now available, and for a small consideration I would be happy to provide you with a list of excellent vendors and public-relations representatives who are fully

Preparing a fossil for discovery.

vetted and qualified to help you make, and then properly promote, any discovery that you desire.

By acting in accordance with the above advice, you will do your share to expand the storehouse of scientific knowledge, lift humanity's intellectual horizons, and bring joy and wonder to untold millions seeking enlightenment.

There is no surer road to fame.

[14]

How to Profit from the Terraforming Program

As the astute reader will, by this time, have observed, we Martians view most of the activities and attitudes of the Mars Authority with a certain amount of ambivalence. However, there is one Mars Authority initiative that we all support 100 percent, and that is their "terraforming" program. Yes, I know that the name reeks of Earthling chauvinism, implying as it does that improving our planet and "reforming" it along Terran lines are equivalent concepts, but peace—we're for it anyway. If they need to be snotty about the program label for reasons of pride, or perhaps to keep selling it to the even worse jerks they report to back on Earth, we can forgive them. Because the terraforming project is the greatest thing that anyone has ever attempted, anywhere, anytime. Period.

Mars was once a warm and wet planet. This fact is obvious to everyone. If you travel around our world, you are beset on all sides by evidence of past water action; dry ponds, lakes, streams, and rivers are to be found everywhere. Take a trip up north, and you can see for yourself the spectacular salt-encrusted shore of what was once our planet's ancient ocean. But other reminders of the past presence of liquid water, including not only salt deposits but sedimentary and conglomerate rocks, are so common that some of them were even discovered by NASA's feeble robotic Mars Exploration Rovers way back at the beginning of the twenty-first century.

Yet today, not a drop of liquid water is to be found anywhere outside of a dome on the Martian surface. The aqua is still here, of course, oceans of it, frozen in the form of ice or permafrost, but except for hydrothermal subsurface reservoirs, all of it is too cold to flow.

Mars's water was liquid in the olden days, because at that time, its carbon dioxide atmosphere was much thicker, and this provided our world with the benefits of a powerful global greenhouse effect. So the planet was made warm enough for an active water cycle, complete with rivers, lakes, oceans, and rain. But when the water rained down through the CO_2 air, it would capture some of the gas in solution, and then react it with soil to form carbonate minerals. This happens on Earth, too, but that planet is so huge that it has not yet lost most of the molten heat it had at the time of its formation. Thus its vast interior reservoirs of geothermal energy still drive the old world's continents all over the place through a process known as plate tectonics. This constantly causes their material to be subducted underground where the heat breaks down the carbonates, thereby allowing the CO_2 they contain to be recycled back into Earth's air. On a normal-size planet like Mars, however, this weird (albeit arguably beneficial) process does not occur, and as a result, once atmospheric carbon dioxide is fixed into carbonate minerals, it stays there.

Thus, over hundreds of millions of years, the thick carbon dioxide blanket warming our planet in its youthful years gradually thinned out, causing the global temperature to drop precipitously, just as happened on Earth after implementation of the disastrous Bali anti-global-warming treaty during the last century. But because Mars is farther from the sun, the results were much worse. Instead of a few decades of blizzards, failed harvests, and brief glacier advances, covering at their maximum extent less than a quarter of North America and Eurasia (which regions, for all the overwrought hand-wringing in the Earthside media, had been comparatively lightly populated in any case), our planet experienced a true catastrophe. On Mars, as the atmospheric CO_2 thinned out through carbonate fixation, temperatures actually dropped to the point where the soil became an effective sorbent for the gas, sponging it right out of the air. And the colder it got, the stronger the soil sorbent became, resulting in a runaway icebox process that froze the whole planet to death.

But in the 3 billion years since that disaster occurred, the sun has grown in power by some 30 percent, so that now it is believed that a further temperature increase of about 10°C could trigger the reverse process. That is, if we today can somehow artificially induce a certain amount of positive global warming, the increased temperature itself would cause some of the CO_2 currently sorbed in the soil to outgas, which would thicken the atmosphere and add to the greenhouse effect. This would warm the planet still more, which would cause yet more CO_2 outgassing, and thus still more warming, and so forth, until the whole planet is as balmy as Tahiti.

Think of it! A new Mars, featuring groves of tall coconut palms fronting an azure sea whose slow moving but triple-height waves carry bevies of beautiful near-nude surfer girls and boys lazily to the shore, where they join you to tan by day or relax by night amidst the warm red sand.

OK, so I may have gotten a little carried away there. Although in my defense I should mention that such projections are to be found in the publications of many highly reputable real-estate firms, and who am I to contradict them? But even if you listen to the sourpuss scientists working with the terraforming project, the picture is still pretty exciting. According to these *pessimists,* if we can kick things off by generating an artificial planetary temperature increase of about 10°C, the positive feedback caused by the release of sorbed CO_2 from the soil should continually amplify this, so that within a century, Mars will have a 200-mb-thick CO_2 atmosphere and an average global temperature about 50°C higher than what prevails today.

Yes, I know; Mars's current global temperature is –55°C (or 218 K), so a 50°C temperature rise would still leave us with a planetary average a bit less than the 0°C freezing point of water. But that's the *planetary average;* New Plymouth is close to the equator, so the temperature around here (and in near-tropical Tsandergrad and Taikojing, for that matter) would be several degrees above the freezing point year-round—and even as high as 40° latitude, liquid water would be possible in the summertime.

So maybe Tahiti is not a precise analog. Perhaps Alaska might be closer to the mark. Yet still, under these conditions, the vast amounts of water frozen into our planet's soil would melt. The long-dry streams and riverbeds of Mars would flow once more, to fill again its ancient lakes and oceans. There would be rain, and snow, and water falling and melting everywhere to break down the peroxides in the soil, detoxifying it globally while adding several mb of oxygen to the atmosphere.

Thus, Mars could be transformed from its current dry and frozen state into a warm and wet planet capable of supporting life. We Martians will not be able to breathe the air of the newly remade Mars, but we will no longer require spacesuits and instead will be able to travel freely in the open wearing ordinary clothes and simple breathing masks. In addition, because the outside atmospheric pressure will have been raised to human-tolerable levels, we will be able to make enormous habitable areas for ourselves under huge domelike inflatable tents containing breathable air. The domes could be of unlimited size because, unlike the pressurized domes we need to use today, there would be no pressure difference between their interior and the outside environment.

But even if we might not be able to breathe the outside air straightaway, simple hardy plants could, and in fact would thrive in the carbon dioxide–rich outside environment to spread rapidly across the planet's surface. In the course of centuries, these plants will introduce oxygen into Mars's atmosphere in increasingly breathable quantities, opening up the surface to advanced plants and growing numbers of animal types. As this occurs, the carbon dioxide content of the atmosphere will be reduced, which would cause the planet to cool unless greenhouse gases are introduced that are capable of blocking off those sections of the infrared spectrum previously protected by carbon dioxide, but even the Mars Authority is smart enough to deal with a matter like that.

It might take a while, but eventually the day will come when the domed tents will no longer be necessary, and our descendants will be able to throw away their oxygen masks to inhale the glorious scent of the towering evergreen forests of Mars.

This is our magnificent vision, the underlying faith shared passionately by every Martian, from the stodgiest bureaucrat in the Mars Authority to the toughest prospector in the outback or the sharpest operator in the spaceport Sisterhood. We are here for a reason, *to bring life to Mars, and Mars to life.* To this cause we commit our lives, fortunes, and sacred honor, and you can bet your last kilo of krill that it is going to happen, because, come what may, we will never give up until we succeed.

That said, the really important question is, how can *you* score a profit from all this?

Making Money from Terraforming

The terraforming program is truly wonderful because, putting aside all the obligatory piffle about holy grandeur and so on and so forth, it affords so many terrific ways to make piles of money.

This wasn't always so. Initially they had this plan created by a bunch of UN boobies who thought they would do the job by dropping the complete arsenal of Terran nuclear weapons on the south polar cap, vaporizing the large amounts of frozen CO_2 stored there, and thus kick-starting the warming process. Apparently they thought this might be a good way to achieve disarmament on Earth and didn't care about the potential damage to real-estate values here caused by all the radioactive fallout. Fortunately, however, the Russo-Iranian nuclear war happened just in time and convinced the practical people in the various Earthside governments that they needed to hold on to their nukes in case they were needed for more appropriate purposes.

Then there was a scatterbrained plan put forth by NASA, who proposed to send some nukey ships out beyond Pluto to the Kuiper Belt to find billion-tonne ammonia-ice asteroids and then give them a shove to destabilize their orbits so that they would come falling into the solar system to collide with Mars, heating it both by impact and via the release of ammonia, which is a strong greenhouse gas. This concept initially seemed attractive to many people, as it offered an excellent potential for sale of impact-bombardment insurance policies. However, market surveys revealed that most of the likely customers who thought their communities might be blasted to smithereens by inaccurately aimed incoming asteroids could not be convinced that insurance policies issued by local providers would be of significant value to them, as the providers would likely be incinerated in the same event. Thus, all the business would go to Terran providers, making the program completely useless to us, especially in view of the facts that NASA didn't know if there really were ammonia-ice asteroids in the Kuiper Belt anyway, and didn't have a clue as to how to deorbit one with sufficient accuracy to make sure that it would hit Mars at all, rather than miss, or hit some other nearby planet, such as, for example, Earth (duh). In addition, there was the problem that, while in principle a nukey ship might be able to reach the Kuiper Belt in a decade or two and, with luck, stop itself there rather than fly off into interstellar space, it would take about a century for a deorbited object to fall into the inner solar system and hit any target at all. This unfortunate fact further served to undermine any likelihood of near-term insurance sales. Despite adamant campaigning for the plan by both the Nuclear Spacecraft Development Office and the Outer Solar System Exploration Office of NASA (which continues to this day), the combined factors that it (a) would not work, (b) might cause mass devastation, and most important, (c) did not offer any real profit potential to anyone outside of the NASA contractor community, ultimately caused it to be rejected.

Thus, the plan ultimately adopted was that proposed by the Mars Authority, guided, as always, by late-twentieth-century thinking. At that time, there was a lot of hysteria about atmospheric releases of chlorofluorocarbon gases, or CFCs, which were considered to be an apocalyptic danger because they cause damage to the Earth's ozone layer, and were thereby threatening to increase the Earth's surface ultraviolet dose from something like 1 percent Mars normal to 2 percent (gosh). In addition, on a molecule for molecule basis, the CFCs were found to function as extremely powerful greenhouse gases—although because the aggregate released was insignificant in comparison to industrial CO_2, that was also immaterial. Nevertheless, since the Terran press of that time were as much a bunch of silly fussbuckets as they are today, this put CFCs in the news and drew them to the attention of the tiny handful of Earthlings who were actually rational, and thus concerned with meaningful issues, like the settlement of Mars, as opposed to the wars, scandals, stage sensations, health plans, tax schemes, corporate swindles, political power struggles, and innumerable other transitory matters that obsessed the rest.

Thus, not long after the Viking probes made clear that Mars had once been warm and wet, and some of these proto-Martian thinkers began to speculate on how it could be made so again, the potential utility of deliberately employing CFC-type gases for such a purpose readily suggested itself. Of course, since Mars, unlike Earth, actually could use some reduction in surface UV levels, ozone-destroying agents wouldn't do, so CFCs as such were ruled out. But after some research, our wise Forerunners hit on the idea of using simple fluorocarbons (such as CF_4, C_2F_6, and C_3F_8) instead of CFCs to do the job, the advantage being that fluorocarbons offer beneficial greenhousing power similar to CFCs, without the ozone-damaging side effect.

Why schoolchildren today are taught that this was a huge intellectual breakthrough is incomprehensible. The idea seems utterly obvious to me. But peace, I don't mean to criticize the Forerunners. I am, after all, a patriot. I guess I am just disgusted that so many people today who claim to admire them actually use their enshrinement as an excuse for failing to effect comparable accomplishments. I mean really, use FCs instead of CFCs; how hard could that have been to think of? Sure, it was something of a step forward in its day, but it's been over a century. You'd think that in the meantime someone in the well-funded Mars Authority Terraforming Directorate might have found an additional small advance that might speed things up? No, of course not. Only the demigods of the past could be expected to do something like that. Pish!

So that's the Mars Authority plan; follow scripture and do it just the way the Forerunners said, by setting up factories to manufacture fluorocarbons, and dump them at a rate of a thousand tonnes an hour into the atmosphere. It may be unimaginative, it may be wasteful compared to more modern methods that certainly could have been discovered if anyone in the Mars Authority Terraforming Directorate (MATD) had any brains, but all the scientists agree that it should work.

And now that the program is finally well under way, dishing out big contracts to everyone in sight, you have a chance to use it to make some serious cash. There are many ways to do so. Consistent with our preference for operating with due respect for statute, we'll start with the legal ones first.

You can, if you provide an appropriate gratuity to the right MATD officials, easily get yourself a substantial contract for the construction or maintenance of one of the many factories being built for producing the fluorocarbon gases that are the centerpiece of the terraforming effort. The problem with construction jobs, however, is that while you can make a lot of money by billing the customer for first-class components while actually using discards, there is too great a chance of being held liable if the place explodes, or even just falls apart due to some relatively benign failure. Maintenance contracts seem like an excellent money train, but you need to realize that the scamp-built factory you will be repairing may well be a death trap.

So rather than get too intimately involved with the MATD industrial facilities themselves, a cleaner way to make a score is to take part in the effort to supply the program with its raw materials. To make fluorocarbon gas, two things are needed: carbon and fluorine. Since CO_2 constitutes the majority of our atmosphere, even the Mars Authority knows how to get carbon for itself, but finding and acquiring fluorine requires knowledge, talent, courage, and a lot of good old-fashioned honest hard work, and so, by necessity, they have had to turn to others. That's where you come in.

Commercially useful fluorine can most readily be found in the fluoride salt deposits that litter the shores of many of the ancient lakes and ponds. In addition, fluorosilicate minerals are sometimes encountered in significant concentrations in the highlands. So, if you are up for a little field exploration, one way to make some serious money is just to go out and find the stuff and stake some claims. This is getting somewhat harder to do than it used to be, since all the really big deposits near New Plymouth have probably already been found, but still, for those daring enough to venture out further than others have so far deemed safe, plenty of treasure still surely awaits. Remember: *Fortune favors the brave.*

Admittedly, however, engaging in such work is a bit of a gamble, since many of those who go where no one has ever returned from don't return themselves, either. As an alternative, you could join or form a consortium to buy existing fluoride salt claims and actually mine them, but the profit margins in such complex businesses are narrow, the headaches are many, and you can actually lose money if you have an excessive rate of equipment failure. However, there is a sure bet for easy cash in terraforming, and that is to get into the business of delivering the fluoride salts from the mines to the MATD.

Yes, I know that sounds like a strange recommendation, since the MATD has established audited transport rates of so much chargeable per tonne-kilometer across various types of terrain, and has cleverly set these rates at precisely such a level that the best anyone can hope to make is about 5 percent above costs on each shipment. However, what they don't necessarily know is where most of your transports actually shipped their goods from. So long as you have a few trucks that you keep deployed near some distant mines so as to register repeated appearances there to take cargo, which can be dumped in the outback, with perhaps an occasional real cross-country delivery done for appearances' sake, you can use the rest of your fleet to bring in stuff from much closer reserves and charge the premium long-distance shipment rate.

Relatively nearby reserves you might consider accessing could come from one of the old local mines, which, contrary to what the MATD thinks, have certainly not tapped out. But the sharp characters who own those places will know your game, and they'll want their cut. So my advice is to not mess around, and just get the bulk of your fluoride salt delivery cargo from the most convenient source of all, which is to say the MATD salt-storage bins to the rear of the factory. The materials stored in these facilities can be obtained at very low cost simply by providing an appropriate gratuity to the MATD security personnel. You can then do a nighttime pickup, or, if you prefer the work to be performed professionally, arrange to have one of the Sisterhoods handle it in return for a modest charge. Either way, your profit margins will almost certainly be excellent, and you will be able to acquire capital to continually expand your transport fleet and make yourself an ever more significant player in this sacred effort that promises so much for the future of life, humanity, and Martian civilization.

However, you don't even have to be directly involved in the terraforming program in order to make money from it. The mere fact that it is ongoing promises to increase real-estate values nearly everywhere on Mars. That said, some places will increase in value much more than others. So all you need to do to make a fortune is to grab the right properties and then resell them to saps who weren't as quick as you to get in on a good thing.

Please note, the fact that most of the effects of the terraforming effort won't actually occur for at least another century is irrelevant. Since everyone knows that extraordinary physical improvements are on the way, market values for selected properties are already taking off, and many more can be made to soar, provided things are handled correctly.

As an important example of the above, consider the potential sales value of future beachfront property. On Earth, properties that front bodies of water sell for a high premium, and the same will obviously be true on Mars once the terraforming program brings back into being our planet's many ancient ponds, lakes, rivers, seas, and oceans.

Now it may be pointed out that on Earth, it is known precisely where the shore of a lake or an ocean actually is, whereas we don't know exactly how high sea level will rise on Mars, so a property that might be a prime future beachfront value with equal likelihood could end up far from shore—or worse yet, underwater. While this may sound like a problem, nothing could be further from the truth. In fact, it opens up huge opportunities, since it means that *any* property on the slope of a basin or valley can be marketed as future beachfront. All that is needed to do so is to obtain an appropriate expert opinion identifying the site in question as being adjacent to the definitive location of the future shoreline. Such opinions, backed up by unquestionable computer calculations, can be readily obtained from many noteworthy and highly credentialed members of the MATD scientific staff, in exchange for a small piece of the action.

Another good opportunity created by the terraforming program lies in the field of hydroelectric power. On Earth, most of the expense involved in building a large hydroelectric dam is a consequence of the huge effort required to divert a flowing river away from the dam site so that the structure can be built, and then divert it back. On Mars, however, we don't have that problem, as the rivers

The terraforming program will transform barren desert land into high-value tropical beachfront property. The time to buy is now.

that will drive our hydroelectric installations do not yet exist. Thus it will be possible to build hydroelectric dams here at much lower cost than was ever possible on Earth. Furthermore, given the extreme variations in topography that Mars affords, with mountains 27 kilometers high and canyons 5 kilometers deep, it is clear that the hydroelectric potential of our planet is immense and the future revenues of our hydroelectric industry will be astronomical.

So all you need to do to get in on this bonanza is to start your own hydroelectric company by buying up a prime future dam site. Nearly any stretch of old runoff channel will do for this purpose, provided you exercise due diligence to get a sufficiently prestigious endorsement of its critical value from one or more distinguished experts drawn from the hydrogeological research staff of the MATD. (One is generally sufficient, as these people tend to have terrific job titles.) Provided that you show adequate respect for the value of their opinions, this can always be arranged. Once it is, you have a stock offering prepared, and then go public with it on the Rangoon or Lagos exchange, where it is sure to be snatched up by the savvy investors who abound in such locations.

In no time at all, you, your MATD scientific advisor, and the rest of your company marketing staff could all be krillionaires. Not only that, you will earn an immense amount of goodwill for yourself, the MATD, and the Martian community in general, because the Earth-side investors who buy your stock will no doubt make terrific profits themselves unloading the paper on others, and those others on still others, for at least a century, as the terraforming program advances and the prospect of actually producing some hydroelectric power becomes ever more tangible, at least in theory. Remember: *Property titles are not for using, they're for buying and selling.* Keep that fundamental truth in mind, and you're certain to make out really well.

I could provide plenty of other examples, but I think you get the idea. Representing as it does humanity's highest hopes and aspirations, the terraforming program is unmatched by any other project in mankind's history in terms of the promise it offers to the bottom line of those who choose to embrace its profound vision.

Life to Mars, and Mars to Life. Yeah brother, amen.

Technical Note (WARNING: High Science Content)
The Science of Terraforming

While the concept of terraforming Mars may seem fantastic, the concepts supporting the notion are straightforward. Chief among them is that of positive feedback, a phenomenon that occurs when the output of a system enhances what is input to the system. For a Mars greenhouse effect, we find a positive feedback in the relationship between atmospheric pressure—its thickness—and atmospheric temperature. Heating Mars will release carbon dioxide from the polar caps and from Martian regolith. The liberated carbon dioxide thickens the atmosphere and boosts its ability to trap heat. Trapping heat increases the surface temperature and, therefore, the amount of carbon dioxide that can be liberated from the ice caps and Martian regolith. And that is the key to terraforming Mars—the warmer it gets, the thicker the atmosphere becomes; and the thicker the atmosphere becomes, the warmer it gets.

To understand how this works, take a look at the graph on the next page, which shows the dynamics of the Martian regolith/atmosphere CO_2 greenhouse system. The curve marked with the little squares shows the average global temperature as a function of the CO_2 atmosphere's pressure. Here we see the predicted results of the greenhouse effect. The thicker the atmosphere, the warmer the planet gets. The line with the diamonds shows the soil vapor pressure as a function of the global temperature. The warmer the planet gets, the more CO_2 vaporizes from the poles and outgases from the regolith.

Note the two points, A and B, where the curves cross. Each is an equilibrium point where Mars's mean atmospheric pressure and average temperature (given in degrees Kelvin—to translate into centigrade, just subtract 273; 273 K = 0°C) given by these two curves are mutually consistent. However, A is a stable equilibrium, while B is unstable. This can be seen by examining the dynamics of the system wherever the two curves do not coincide. Whenever the temperature curve lies above the vapor-pressure curve, the system will move to the right, or toward increased temperature and pressure; this would represent a runaway greenhouse effect. Whenever the temperature curve lies below the pressure curve, the system will move to the left, or toward

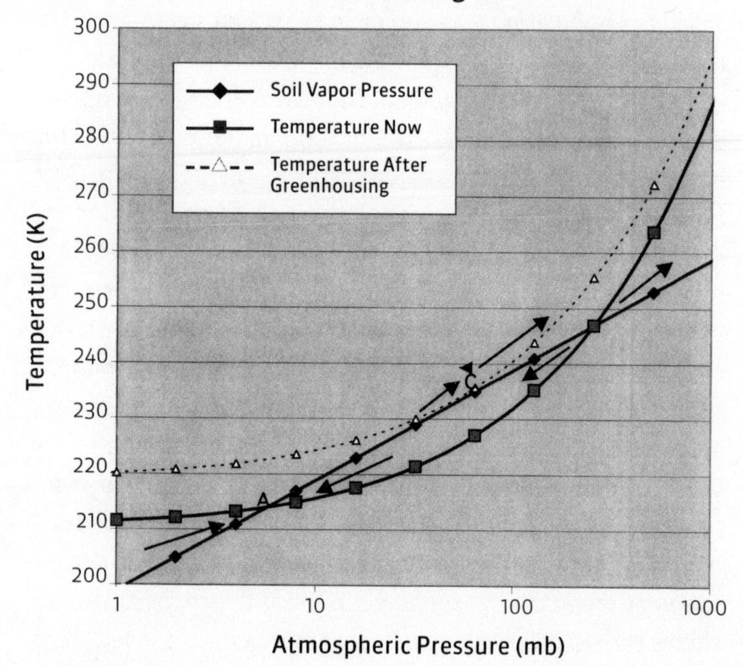

To terraform Mars, just lift the temperature curve above the soil vapor pressure line. Once points A and B meet at C, the system will have no stable equilibrium and a runaway greenhouse effect will result.

decreased temperatures and pressure; this would represent a runaway icebox effect. Mars today is at point A, with 6 millibar of pressure and an average global temperature of about 215 K.

Now consider what would happen if someone artificially increased the temperature of the Martian poles by 8 Kelvin. The results of such a change are shown by the dashed curve marked with little triangles. As the temperature is increased, the solid temperature curve would move upward to where the dashed curve is marked, causing points A and B to move toward each other until they meet at point C. It will be observed that the average global temperature at C is 230 K, which is 15 degrees warmer than the 215 K temperature we started with at A, so it is clear that the effect of the original artificial 8 degree temperature rise that was put into the system has been substantially amplified by positive feedback. But more important, the new temperature curve is above the pressure curve everywhere, so point C is an unstable equilibrium. Once this is reached, the result would be a runaway greenhouse effect that would cause the entire available reservoir of polar and regolith CO_2 to evaporate and outgas, driving temperatures and pressures upward and outward along the dashed curve. As soon as the pressure has moved out past the current unstable equilibrium location, the roughly 200 mb point B, Mars will be in a runaway greenhouse condition even without artificial heating, so even if we stop the heating activity later, the atmosphere will still remain in place.

With 6 mb of CO_2 in the atmosphere now, close to 100 mb frozen at the poles, and perhaps 400 mb in the regolith, there is probably enough available CO_2 to create an atmosphere with around half the surface pressure of Earth. Looking at the data on the graph, we can see that under such conditions the average global temperatures might rise to about 275 K, i.e., slightly above the freezing point of water, with tropical equatorial and summertime temperate zone climates being considerably warmer.

That's good enough to create a living planet.

How Fast Would the Atmosphere Come Out of the Regolith?

But how fast could this be done? The stuff obtained from the polar cap will come off quickly, but forcing out adsorbed carbon dioxide from regolith at significant depth might take some time. For terraforming to be of practical interest to investors seeking a *high rate of appreciation on their real-estate claims, the rate at which all this occurs is important. After all, if it takes ten million years for a substantial amount of gas to come out of the regolith, and potential land purchasers find out that this is likely to be so, the fact that it comes out eventually would be rather academic.*

Fortunately, in this case, there is no need to pay extra premiums to scientific charlatans to improve the facts. The rate at which gas comes out of the regolith will be in direct proportion to the rate at which a temperature increase that we create on the surface of Mars can penetrate into the ground. The thermal conductivity of Martian regolith is a lot like dry soil on Earth, with maybe a little bit of ice mixed in. The rate at which heat will spread through such a medium will be governed by the process of thermal conduction, whose equations predict that the time a temperature rise needs to travel a given distance through a medium is proportional to the square of the distance. The MATD scientists have measured this rate at various places on Mars, and their average works out to about 16 square meters per year. Martian regolith, which has an average density of about 2.5 tonnes per cubic meter, includes a lot of claylike minerals, and the best guess of the top docs at the MATD is that it is saturated with about 5 percent carbon dioxide down to considerable depth. If this is true (and who am I to dispute the conclusions of the well-credentialed scientists of the MATD?), we would have to force out carbon dioxide held in regolith—outgas it—down to a depth of 100 meters to produce a 500 mb (half Earth sea-level) pressure on Mars. So let's say we induced a sustained artificial temperature rise at the surface of 10 K, good enough to outgas a significant fraction of what is in the regolith. This temperature rise would then travel down into the ground. The rate at which this would occur is shown in the graph on page 161.

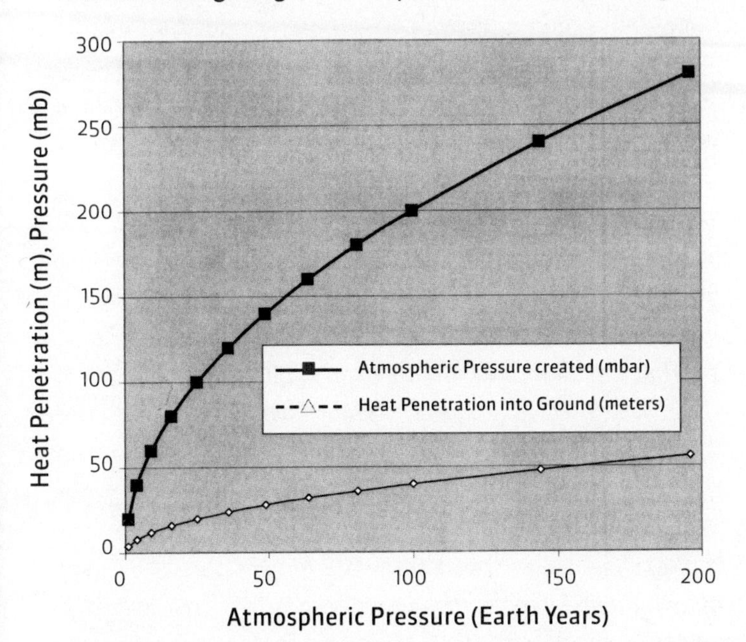

After the surface warms, it will take time for the heat to soak into the ground. In 100 years, the warming will reach a depth of 40 meters, releasing 200 mb of CO_2 into the atmosphere.

You can see that while it takes a very long time to reach significant depths, modest depths can be reached rather quickly. So, while it might take 200 years to penetrate 100 meters to get about 300 mb out of the regolith, the first 100 mb can be gotten out in just a few decades.

Once significant regions of Mars rise above the freezing point of water on at least a seasonal basis, the large amounts of water frozen into the regolith as permafrost would begin to melt and eventually flow out into the dry riverbeds of Mars. Water vapor is also a very effective greenhouse gas, and since the vapor pressure of water on Mars would rise enormously under such *circumstances, the reappearance of liquid water on the Martian surface would add to the avalanche of self-accelerating effects, all contributing toward the rapid warming of Mars. The seasonal availability of liquid water will also allow us to spread bacteria, which will produce methane and ammonia that will augment the greenhouse effect and also protect the planet against solar ultraviolet radiation, as well as green plants, which will begin the process of oxygenating the atmosphere.*

So, in short, the science says that if we can somehow raise the planet's temperature by 10°C or so, we can make our world come alive. That's it; all we have to do is induce 10°C of global warming, and nature will take care of the rest. But how do we do that?

Producing Halocarbons on Mars

The most obvious way to increase the temperature on Mars is simply to set up factories to produce halocarbons, which are the strongest greenhouse gases known to man. In fact, one variety of halocarbon, known as chlorofluorocarbons, or CFCs, had to be banned on Earth in the 1990s because of its strong contribution to the greenhouse effect, and because it was blamed for the destruction of the ozone layer. However, by choosing our halocarbon greenhouse gases carefully to employ varieties lacking chlorine (i.e., fluorocarbons or FCs), we can actually build up an ultraviolet-shielding ozone layer in the Martian atmosphere. The easiest such gas to make is perfluoromethane, CF_4, which also has the desirable feature of being very long-lived (stable for more than 10,000 years) in our planet's upper atmosphere. The greenhousing effect of using CF_4 can be improved by throwing in smaller amounts of other fluorocarbon gases, such as C_2F_6 and C_3F_8, which serve to block up the gaps in the infrared spectrum that an atmospheric blanket of CF_4 and CO_2 alone might leave open. In table 1 you can see the amount of such a fluorocarbon gas cocktail needed in Mars's atmosphere to create a given temperature rise, and the power that we would need to generate on the Martian surface to produce the required fluorocarbons over a period of twenty years. If the gases have an atmospheric lifetime of one hundred years, then approximately one-fifth the power levels shown in the table will be needed to maintain the FC concentration after it has been built up. As you can see, we are going to need a pretty substantial industrial effort to pull this off—something like 2 to 4 gigawatts (a gigawatt, GW, is 1,000 megawatts) if we are going to build up a gas blanket in a timely way. This would not be much for Earth, where a gigawatt is wasted just to provide the power to a typical no-name American city in the 1-million population class, but it is nearly the entire amount of power we currently have available planetwide on Mars. So it's going to take a while before we can build up our power capacity to really put this program into high gear, but that is no reason not to profit by selling high-value land based on its assured future success today.

Table 1: Greenhousing Mars with FCs

Induced Heating (degrees K)	**FC Pressure** (microbar)	**FC Production** (tonnes/hour)	**Power Required** (GW)
5	0.012	260	1.31
10	0.04	880	4.49
20	0.11	2,410	12.07
30	0.22	4,830	24.15
40	0.80	17,570	87.85

(Data courtesy of the MATD)

Oxygenating the Planet

As the planet warms, its hydrosphere will be activated. Water will melt out from the ice and permafrost, flow into the streams, rivers, and lakes, evaporate, and come down everywhere as rain and snow. The more rapidly water gets into circulation, the more the action of denitrifying bacteria will break down nitrate beds and increase the atmospheric nitrogen supply, and the spread of plants to produce oxygen will be accelerated. Activating the hydrosphere will also serve to destroy the oxidizing chemicals in the Martian regolith, thereby releasing some additional oxygen into the atmosphere in the process. But releasing enough oxygen into the air to make it breathable for us is going to be challenging. Bacteria and primitive plants can survive in an atmosphere without oxygen, but advanced plants require at least 1 mb and humans need 120 mb. While Mars does have superoxides in its regolith and nitrates that can be heated to release oxygen and nitrogen gas, going about things that way would require enormous amounts of energy, about 2 million gigawatt-years for every millibar produced—and that's just too expensive to be practical unless somehow we can con the Earthlings into paying for it.

Similar amounts of energy are required for plants to release oxygen from carbon dioxide. Plants, however, offer the advantage that, once established, they can propagate themselves. The production of an oxygen atmosphere on Mars will thus break down into two phases. In the first phase, pioneering cyanobacteria and primitive plants will be employed to produce sufficient oxygen (about 1 mb) to allow advanced plants to propagate across Mars. Once an initial supply of oxygen is available, and with a temperate climate, a thickened carbon dioxide atmosphere to supply pressure and greatly reduce the space radiation dose, and a good deal of water in circulation, plants that have been genetically engineered to tolerate Martian regoliths and to perform photosynthesis at high efficiency will be released together with their bacterial symbiotes. Assuming that global coverage could be achieved in a few decades and that such plants could be engineered to be 1 percent efficient (rather high, but not unheard of among terrestrial plants), then they would represent an equivalent oxygen-producing power source of about 200,000 GW. Using such biological systems, the required 120 mb

Scientific projection of the future Mars, after terraforming. Note the extensive amount of shoreline property.

of oxygen needed to support humans and other advanced animals in the open could be produced in about 1,200 years.

Yes, I know, that's too slow for most people's taste. But once we engineer more powerful artificial-energy sources or still more efficient plants (or perhaps truly artificial self-replicating photosynthetic machines), then we will be able to accelerate this schedule radically.

I know we can do it. With so much money at stake, Martian ingenuity can't possibly fail. And consider this: the development of thermonuclear fusion power on the scale required for the acceleration of our terraforming project would also create the key technology for enabling piloted interstellar flight. Think about that: we're not just doing this to make ourselves rich. We're giving humanity the stars.

> **Witness this new-made World, another Heav'n**
> **From Heaven Gate not farr, founded in view**
> **On the clear Hyaline, the Glassie Sea;**
> **Of amplitude almost immense, with Starr's**
> **Numerous, and every Starr perhaps a World**
> **Of destined habitation.**
>
> —John Milton, *Paradise Lost*

So there.

Let there be Life!

[15]

How to Be a Social Success on Mars

Now that I have shown you the way to the four principal goals of human life, to wit, how to survive, gain riches, become famous, and play a part in a transcendent cause, it behooves me to spare a few lines to discuss secondary matters that nevertheless remain of interest to some readers. These generally fall under the category of achieving "social success."

Can you find your soul mate on Mars? Someone for you to love, but more, someone worthy of your love, who will love you passionately in return, and stand by you, as your loyal and true comrade, lover, friend, and defender, sharing all joys and facing all dangers and adversity together, eternally and forever, through dust storm and solar flare, through score and scandal, for richer or (heavens forbid) poorer, till death do you part?

The idea is not as silly as it sounds. In the first place, it should be clear that finding such a companion would be of great utility, as he or she could potentially provide you with (a) regular sex, (b) someone to watch your back, (c) someone to lend you money on reasonable terms to get started again if your luck turns sour and you go broke, (d) valuable business connections, and (e) children of a sort who might prove useful at some time in the future. In the second place, contrary to your Earth experience and its derived cynicism, on Mars, such things are possible. Yes, fully possible, even for you, a person who obviously was a complete social failure on Earth—otherwise, you wouldn't be here. You just need to listen to my advice.

But before I begin my discourse on how to find, wisely choose, successfully woo, and partner with a mate on Mars, I need to discuss that one fact concerning our social life that inevitably startles and amazes all new Earthling immigrants once they get over their incredulity: On Mars the institution of marriage still exists. I am not making this up. If you don't believe me, just ask the kids playing around Founders' Square or in the New Plymouth Central Agricultural Dome. Nine out of ten of them have *two* parents, and not only that, nearly all of those have had the *same* two parents for their entire lives. The same pattern holds true in Tsandergrad and Taikojing, as well. I know it must sound unbelievably prehistoric to you, but it is a fact; on Mars, people can and do get married, just as they do in Shakespeare's plays when performed in their unauthorized nonupdated versions.

But perhaps you shouldn't be so shocked. Marriage was, after all, still fairly common on Earth in at least some out-of-the-way places like Lapland, Outer Mongolia, Tierra del Fuego, and southern Utah as recently as fifty years ago, and was normative, if declining, nearly everywhere fifty years before that. The collapse of marriage as a major social institution on Earth is really a comparatively recent phenomenon, brought about by the bureaucracy's mad assortment of divorce laws, domestic-violence laws, child-abuse laws, spousal-certification laws, parental-suitability laws, education laws, indoctrination laws, anti-indoctrination laws, health laws, dietary laws, pharmaceutical-treatment laws, mental-hygiene laws, therapy laws, counseling laws, psycho-certification laws, household-inspection laws, and endless other state invasions that more or less made stable families impossible, at the same time that the general spread of deranged ideas about the supposed evil that humans represented to Nature made their primary previous purpose appear undesirable. Think about that: if you had been born on Earth a century ago, you might well have known (with as much as 50 percent accuracy) who your biological father was, and *been virtually certain as to the identity of your mother!* It's true that they did have state orphanages, but being raised in one was rather the exception than the rule that it is on Earth today. In fact, in premodern English, the very word "orphan" was not a general synonym for "child," but only used to refer to a child whose parents were both dead. If your parents were alive, it was not only legal, but *expected,* that they would raise you *themselves.*

Well, that is how it is on Mars today. There are no state orphanages. Children are born, generally speaking, to married couples who raise them at home, rather than by single women who produce them by accident or as part of their two-birth obligation to the Social Security system. Incredible as it sounds, *people on Mars actually want to have children of their own,* and they form families for that purpose.

Thus, while sexual attractiveness is a factor among us while seeking pairings, unlike Earth, it is not the only factor. If you wish to succeed in the dating game here, you need to understand this, because when you approach someone of the opposite sex, he or she will not just be evaluating you shallowly on the basis of your looks, but will be asking themselves deeper questions like: "Do I want this man to be the father of my children?" or "Do I want this woman to be the mother of my children?" If you want to be able to score with any frequency, it is essential that, as they search their consciences and feelings, your marks are able to answer these questions in the affirmative.

On Mars, families still exist.

Thus it should be clear that the key to making it with members of the opposite sex on Mars is to be rich. Put simply, if you have piles of money, you will easily be able to secure whatever liaison you want, including, but not limited to, the life partner of your dreams. As the wise old saying goes: *Make yourself rich, and love will follow.* So, since we have already dealt adequately with the critical subject of how to get rich, we may also regard the secondary matter of winning the heart you desire as being settled as well.

Nevertheless, that still leaves unanswered the critical question: Whom should you choose to marry? The obvious answer, of course, is others who are rich as well. The problem, however, with this method is that rich mates are very much in demand, and so they tend to get taken off the marriage market so quickly that you may never get a shot at one. Or if they are available, it could well be due to the fact that they are so totally disagreeable in their physical or emotional attributes that no one will take up with them despite the fact that they are loaded. It is therefore the unfortunate case that you may well be faced with the unpalatable prospect of choosing your mate from among the unrich. There is no need to panic, however, since all you need to do to come out well is to use your keen eye for character to identify someone who, while disgustingly poor today, is likely to become delightfully rich in the future. Essentially, that's all there is to it. (If you are unsure of someone, ask them some test questions based on the financial success chapters of this book. If they're too dumb to get it, drop them.)

In addition to obvious future financial failures, there are several other types you need to avoid. These include most of those employed by the Mars Authority, especially any of their hordes of inspectors, regulators, revenuers, and shrinks. It is true that members of the first three categories can occasionally be useful to have in your corner, but having one of them around will make you so odious to all other right-thinking Martians that you will find the association a net debit regardless. As for the fourth, they are useless on all occasions, and are generally quite irritating besides. I mean really, do you want to face a life filled with conversations that go like this:

"Honey, great news. I think I just struck it rich!"

"That's wonderful dear, but how do you *feel* about it?"

Or:

"Did you hear it? Those MA bastards just put a tax on rover repairs."

"I see. So, would you say you are feeling anger right now?"

Get the picture? Yick. Besides, they're incapable of making any serious money.

On the other hand, having a well-qualified Mars Authority geologist in the family can be a real asset, as you can use him or her to validate your mining claims. The two of you will be a popular couple with other Martians for the same reason. Just be sure not to take the same name after marriage, as doing so could cause unnecessary concern among those doing due diligence on behalf of potential purchasers of stock in your bonanza.

If you are a man, you may be tempted to marry a member of the Sisterhood, as doing so will indeed provide you with links to a whole web of valuable connections. But don't do it—the power within the marriage will be too unequal. Marrying a member of the Sisterhood is putting yourself into the same position that husbands were in during the last days of marriage on Earth, when their wives could dispossess them simply by calling 9 on their telephones (shortened from 911, to make the process quicker). She may set you up to score some easy money, but ask yourself this question: Who gets the bird, the hunter or the dog?

Where to Look for a Mate

The kind of person you want for a mate is a man or woman like yourself, someone of real character, a rugged individual, willing to stand on his or her own two feet and make it through honest hard work, courage, and skill. You came to Mars because you are such a person; others have done the same, so they are here to be found. But the best place to find them is not in the bureaucracy or the spaceport syndicates but among those willing to take on the challenge of the new frontier at its front—among the prospectors and miners and daring peddlers of the outback.

So one obvious way to meet your soul mate is just to sign on with a prospecting team and get to know the other people at your camp. This procedure has the great advantage that you really will get to *know* them, and get to see their characters and business acumen tested in action. As a way of choosing a true comrade for your lifelong partner, this can't be beat.

Unfortunately, however, a typical prospecting camp might only have a dozen people, and so the number of available eligible candidates to choose from in yours could well be limited, or nonexistent. So the place you are most likely to find the prospector boy or peddler girl of your dreams is not among those working your own small camp but at the grand jamborees where the outback people come together for a rousing good time with outdoor rover races and in-dome dances that show one and all that we Martians really know how to have *fun*.

You can meet your soul mate on Mars, even while working. Just stay alert.

The Pickup

Very well. So let's say you are at one of these events, or are merely standing in line at a film revival in New Plymouth, and you see someone who strikes your fancy. How can you initiate a conversation? There is no universal rule, of course, but here are some standard openers that have stood the test of time:

- Hold still! I think the oxygen line on your suit is loose! Let me tighten it for you.
- Wait a minute. You better get that rover checked before you take it out. So, do you need a ride somewhere? I'll drive you in mine.
- You look like the nurse who debriefed me when I had my debarkation medical inspection. No? Well, it's not too late . . .
- Cold today, isn't it?
- New in town? Need a hab for the night?
- Hi, I'm looking for a prospecting partner too. But don't you think we ought to get to know each other first?
- Did you hear they are expecting an aurora tonight? The best viewing will be a hundred kilometers to the north, but I've got a pressurized rover that can get us there.
- Have you heard about the new two-person-model cocoon bag they are selling at S&R? It's really something. I just got one. Want to see it?
- Hi. I'm looking for a lost baby goat. I think it ran into your hab. Can we go look for it?
- Are you really as beautiful as I imagine you are under that suit?

- Excuse me, but didn't you used to be in movies on Earth?
- Hi. Do you need a partner to practice cocoon emergency procedures?
- Hi, do you need a sponge-bath partner?
- So, the word is we need to conserve on nighttime heating power. Have you thought about how you could help?
- Have you ever had tilapia for breakfast?
- Hi, do you need a place to go to get out of your suit for a while?
- Want to see my rock collection?
- Do you need a physical exam? I do them for free.
- Excuse me, but didn't you used to be a skinsuit model?
- I heard that they got a new case of banned films off the last ship from Earth. Do you know the show schedule?
- It's kind of noisy here, don't you think? I have some homemade hooch and some new recordings at my hab. Why don't we go there and talk?
- Do you like mushrooms?
- You certainly are lively on the dance floor. How are you at home?
- Hi, I'm new in town. Can you give me directions to your hab?
- Don't tell me you are from the old U.S. of A. Really? So am I.
- Is that a Russian accent? You must like poetry.
- Excuse me, but are you paramagnetic?

- Didn't I know you on Earth?
- Wait a second. I think there is something on your faceplate.
- I have a fantastic book for you to read, but I don't dare let it out of my hab.
- Where did you get that great pressurized rover? Care to take it on an overnight trip to check out some possible alien artifacts?
- Is the claim office near here? I'm new, and I've gotten all turned around . . .
- I know you from somewhere. . . . Did you immigrate from Earth?
- Did I get on the wrong ship? I thought I was going to Mars, but this must be heaven.
- Hi. I'm looking for a guy to help out on my salvage team, but I need to test your muscles first.
- Wow! Have you seen what a beautiful job they have done on the dome garden?
- Do you have a wrench? Are you good with it?
- What in the world is that you are drinking? Did you make the catalyst to synthesize it yourself? Could we go test it in my still?

Where to Go to Have Fun

People like to have fun. So, if you find yourself courting a prospect whose net worth places him or her significantly beyond your just deserts, you can sometimes improve your apparent marriage value enough to seal the deal regardless by showing that you know how to have a really good time. Thus the importance of knowing the best places on Mars to go to have fun can hardly be overstated.

Within the settlements, the liveliest public place for merry mingling is the dance hall. Mars is free of two-thirds of the gravity field that burdens Earth, and so the superior levity of our environment has brought dancing into a whole "new world" of possibilities. Dances like the jitterbug, which went out of style on Earth when men became too weak to lift women readily, are especially popular, as both partners toss each other about with frenetic gaiety, and perform wonderful tricks together in the air. But the waltz can become something even more extraordinary in one-third gravity (provided you have the right partner), as those moments of special intimacy that can occur during a closely held gliding flight are simply not to be missed (especially if you show some initiative, and follow up appropriately afterwards.)

For lighter amusement, there is the large full-Earth-atmosphere dome that the Mars Authority maintains for its senior personnel just outside of New Plymouth. If you sneak into this place with your honey, the thick air there will allow the two of you to strap wings on your arms and soar like birds in Mars's one-third gravity. The fact that this upsets the bureaucrats in residence there to no end only adds to the fun. You need to watch out for the flocks of airborne chickens, however, which have multiplied out of control. If your date is a neat freak, you would be wise to keep your flight altitude well above that generally frequented by the hens. Otherwise, though, engaging your aerial skills to herd the flocks over suitable ground targets can be great sport, whether done competitively or as a team effort.

If you can afford an excursion in a ballistic hopper, the place to go is Olympus Mons. This extinct volcano is 27 km high, making it three times as tall as Mount Everest on Earth. Take the hopper to the summit, and then strap roller skis onto your boots for a trip down that the two of you will never forget.

For a good time, take your date flying in the full-Earth-atmosphere Mars Authority headquarters dome. Note: Be sure to stay above the chickens.

Another enjoyable hopper excursion destination is the north polar cap, which is covered with water ice. They have sleds at the base there, which they tie to tethered balloons. These capture the high-speed winds aloft, which then whip the sled along at speeds of over 100 km/h. Trust me, it's quite a ride.

Finally, for those who prefer to save their romantic adrenaline for more intimate moments, there is nothing like the dirigible tour of the Valles Marineris. This canyon is 5 km deep and 3,000 km long, and the towering majesty of the incredible cliffs that parade by you as your airship cruises slowly down the slot inevitably creates a mood that, if you can afford a good stateroom, is simply not to be missed.

These out of town excursions *are* expensive, but the object of your desire will certainly be impressed. Think of them as business investments. If s/he is worth it, s/he is worth it. If not, see if you can get your date to pay. After all, there is no telling who else you might meet while out traveling.

Educating Your Children

Once you get married and have children, it is of critical importance that you do everything you can to provide them with the best possible education. The identity of the optimal school in your area will generally be well known, but if you have any doubt, it can be readily determined by obtaining full-disclosure data on the income distribution of the parents of the attending pupils. By enrolling your children in a school populated principally by the scions of wealthy families, you will maximize the number of valuable business connections that both you and your children can extract from their educational experience, and vastly improve their marriage prospects as well. The benefits of such preparation for success in life can hardly be overstated.

Because the value of such a high-quality education is so clear, Martians have taken definitive steps to create institutions that can assure its delivery. One technique is to form an academy that charges tuition beyond the reach of all but the well-to-do. This definitely does the job, but sending your kid to such a school is a real gamble, since the large tuition expenditure has no resale value. For this reason, a more popular plan is to join a community that only issues permits for first-class, brand-new habs with three or more decks. The community then can set up a school with virtually no tuition charge at all, but still keep the poor out simply by maintaining a residency requirement for admission. It is true that you will generally have to lay out a bundle to buy your way into such a settlement, but you get to live in swank quarters for the duration, and when your kids are done with school, you can sell out and recoup your entire cash investment, or possibly more, should local property appreciate. Thus, using such a strategy eliminates virtually all the risk otherwise associated with a high-priced education.

If you can't afford either an elite academy or a move into a wealthy settlement, your best alternative is just to keep your children at home and educate them yourself. Take them out with you prospecting, building, or exploring; teach them how to fix machines, grow plants, heal animals, analyze mineral samples, or plan field expeditions. It's true they won't make any good social connec-

If your child is well-behaved, home schooling is a good low-budget option.

tions with rich kids this way, but at least they'll avoid getting mixed up with poor ones, and, as partial compensation for losing the benefits that come from good schooling, they might actually learn something useful.

How to Achieve Political Success

Once you become wealthy, famous, and well-settled with an attractive, socially and financially useful mate, situated in a good community, with your fine brood of children properly directed on the road to success, the time may come when you might consider enhancing your social position further through the acquisition of elective office.

The ongoing development of political institutions on Mars is a complex subject, of which I will have more to say in the next chapter. Suffice it at this point to say that, while our governmental structure is still very much a work in progress, there are already quite a few public offices open and available for the taking by anyone with sufficient time, connections, and resources to muster the necessary votes. Provided you choose wisely, you can do very well by taking appropriate advantage of these opportunities.

Public powers break down into two types: those that *punish* and those that *reward*. The great Greek philosopher Plato has one of the characters in his celebrated dialogue *The Republic* say somewhere that the happiest of men is he who has unlimited power to reward his friends and punish his enemies, and this may well be true. However, in our modern systems of divided governmental power, no one person can ever hope to attain this blissful state, and so, should you enter politics, you will need to choose which of these noble professions will be your specialty.

Choosing a punishment office, such as that of a prosecutor, can be very gratifying, as you get to destroy people whom you hate. It can also serve as a tremendous source of self-esteem, by allowing you to engage in a crusade on behalf of your favorite abstract principle, crushing those reprobates who dare to demean its importance—and by implication yours—in the process. This is all great fun. Yet one must question the priorities of those who choose such a course for their lives. For while it is true that you can sometimes accomplish some real substantial good by engaging in such activities (for example by putting a difficult competitor out of business), sooner or later the number of enemies you create by such antics will become so great that your downfall will be certain. Think of Shakespeare's Richard III: a high and mighty terror one day, offering his kingdom for a horse the next, and then, to cap it all, slandered as a hunchback the day after that. You don't want that to happen to you. Thus, as tempting as they might be, punishment offices should generally be shunned.

Reward offices, however, are another matter entirely. If you can install yourself in an office where your role is dishing out wealth to appropriate recipients (which you can do with a clear conscience, because the money involved is not your own), you can make innumerable friends of the most useful sort, without incurring the overhead of an excessive revenge threat. It's true that you will still displease some by rewarding others, but the animosity of such disappointed types will generally not be so great that they will actually devote themselves to your destruction. This is especially true, not merely because of the modest nature of their injury, but because they will know that, as a caring dispenser of goodness, you have it in your power to utterly squash them simply by showing your generosity to a sufficiently grateful prosecutor. Thus it is much better in every way to be one who uses his power to show love to his fellow man rather than hate.

Now the Mars Authority has funds to spread around, of course, but the freedom of those with budgetary authority to spend them in truly productive ways is painfully constrained by various levels of bureaucratic oversight. Furthermore, such offices are not open to election, and regardless, as a person with initiative, character, and a clear idea of where the real money is to be made, you don't want to work for the MA, anyway.

The Martian colonies are beginning to create their own representative institutions to manage necessary community functions. These tend to be rather stingy with their money, however, since it comes from the settlers themselves, and some of these guys seem to have nothing better to do with their time than watch the public till and criticize how it is spent. So there is really no point in running for such public offices as they offer.

So, if you want to accomplish anything worthwhile by venturing into electoral activity, you need to target offices that both possess sufficient resources and offer adequate free scope for your talent to make good use of them. Fortunately, there are a growing number of such opportunities, with the best so far to be found among the important public-private partnerships such as the Interregional Highway Commission (IHC).

Mars today possesses few roads, and no highways. Yet it is obvious that in the future, a fully developed highway system will be necessary to promote the continued growth of our planet's economy. Accordingly, a number of Mars's leading citizens have acted to meet this need in a timely fashion by setting up the IHC. Richly funded as it is through bonds floated among Earthside investors who have no need to know how their money is being spent, the IHC provides its elected board members with innumerable opportunities to do good things for Martians of all walks of life—but especially the truly deserving—all over the planet.

Indeed, the IHC program has proven so successful that we are now moving to set up sister organizations to deal with our pressing needs for bridges and railways. The Martian terrain is riddled with canyons, and given the fact that there will someday be roads and highways leading to them, there is no reason why we should not act immediately to put bridges across them as well. Indeed, since the roads will inevitably go to where the bridges are, those who have the wisdom to act preemptively and build the bridges in advance also will be able to secure the most valuable land rights needed by the highway ahead of time. (Look, I don't care whether this pitch sounds believable to you; they're believing it on the Lagos bond exchange, and that's what counts.) The need for a railway system is even more clear, and given the very large amounts of capital that must and can be mobilized for railway construction, even more lucrative. Thus practically everyone on the planet is extremely excited about the birth of our new Interregional Railway Commission.

Mark my words: the New Plymouth–Tsandergrad transcontinental railway project is going to be one of the wonders of the solar system. All you need to do to be among those visionaries *who get to decide how the program money is spent* is line up enough votes to get yourself elected to the board of the IRC. And with so much loot available to pass around after you achieve office, mortgaging the required electoral support beforehand should not be a problem. Of course, as a relatively new immigrant, people might not know you well enough to trust that you will deliver on your promises, so you may have to settle for lesser offices first. But providing you show that you are honest and reward your followers adequately, your popularity will grow, and you are certain to rise from one office to the next, doing well by doing good.

Public service calls.

[16]

How to Avoid Bureaucratic Persecution

Human beings like to be wealthy, and historically, one of the easiest, simplest, and quickest ways to acquire wealth has been to seize the possessions of others. For this reason, since time immemorial, wise men and women have come together to form institutions known as *governments,* capable of assembling and organizing the resources, information, and muscle necessary for the effective pursuit of this important societal objective. As civilization advanced, it became clear that governments could also help fulfill other valuable social functions as well, most particularly that of providing the armed force required to allow those possessing property to hold on to it in the face of the greed and jealousy of the poor.

Such are the legitimate tasks for which government was created, and to which it was limited during mankind's Golden Age. A problem arose, however, when those entrusted with the responsibilities and thus the powers of government decided to go into business for themselves. Thus was born the plague known as *bureaucracy,* which has persecuted and abused humanity ever since.

On Earth today, bureaucracy reigns supreme, and all who live there must submit to its arbitrary dictates. Their ordinances cover nearly all aspects of life, including what is permitted to be made, said, taught, sold, bought, owned, or built, and where, when, how, and by whom it can be done. If any of these limits (many of which are secret or otherwise so convoluted in their formulation as to make them unfathomable) are transgressed, the "offender" is required to pay a ransom or face either imprisonment, dispossession, loss of work permits, business licenses, citizenship, or other "privileges," disgrace, execution, or any such penalty as the bureaucracy finds convenient to generate sufficient terror to insure prompt payment. In addition, regardless of what they do, all people are required to pay a yearly ransom, called a "tax," simply in exchange for permission to exist. If they fail to do so, they are considered to be an offender and treated accordingly. The vast treasure collected from all these ransoms is then used by the bureaucracy to reward itself, as well as pay the large array of liars, charlatans, propagandists, spies, inquisitors, lawyers, clerks, judges, thugs, shrinks, and jailers it needs to employ to enforce its tyranny.

Bureaucracy, alas, also exists on our planet, most notably in the form of the Mars Authority. The creeps who run that organization will do nearly anything in their power to make your life miserable, or impossible, all the while pompously claiming that they are merely acting for the public good. They will, for example, seek to interfere with your business by launching absurd investigations into the validity of your mining claims or scientific discoveries. They will seek proof that items you produce, which obviously work perfectly well, meet various silly specifications that they will allege exist in some obscure records somewhere. They will question whether or how you have come to own the stuff that you do, demanding documentation that you have come by it "legally" (whatever that is supposed to mean). They will subject you to "safety inspections" and then threaten you with all sorts of penalties for noncompliance with any number of putative regulations that no one has ever heard of and that make no sense whatsoever.

All of these pressure tactics are, of course, simply their covert methods for extorting money. In some cases, however, they will even attempt to openly claim that you owe them a "tax." But whether covert or overt, it is essential that you draw the line against these vile stratagems. Because if you fail to do so, you will not only leave yourself open to be looted at whim but contribute to the cancerous growth of an evil tyrannical force whose unconstrained ambition will never be satisfied until it has utterly enslaved every human being on our planet.

Now don't get me wrong. It is necessary to be practical. You can't just turn them down flat every time. That would get them really mad, and they would subject you to no end of harassment. You need to choose your battles—but then again, so do they. Therefore, to minimize conflict, the procedure that has been worked out is that when an MA official brings a possible violation to your attention, you should thank him for his concern with an appropriate gratuity, after which the two of you can move on. Handling matters in this way is quick, efficient, relatively inexpensive, and serves to provide positive reinforcement to train MA personnel to engage in behaviors with positive outcomes. Properly employed, this technique can allow you to avoid 99 percent of the problems you might otherwise have to face from the bureaucracy.

It sometimes happens, however, that you will be confronted with an MA bureaucrat who is completely irrational, or who is convinced that he simply must make an example by punishing you in order to maintain his credibility to extort money from others. In such instances, gratuities will not suffice, and you will need to use the law to defend yourself.

I know the idea sounds ridiculous, but it is not impossible. It only seems ridiculous to you because you come from Earth, a planet that lacks a functional legal system. Thus, for example, on Earth you might be subject to a random "audit" by the bureaucracy's ransom collection agency, which will inevitably determine that the ransom you paid in some previous year was inadequate and that said underpayment was clearly the result of criminal intent on your part. The fact that the ransom amount in question may have been set in consultation with the agency itself will be deemed irrelevant, because according to them, the responsibility for guessing the "right" amount (which is subject to retroactive change, in any case) rests solely with you. Having thus been summarily convicted by the collection agency, you will be fined, and if you are so foolish as to inconvenience them by appealing your case to another part of the bureaucracy (an absurd step, since they all get their take from the same ransom-collection pot), they will publicly disgrace you and possibly imprison you too. (These methods have proven very effective not only against ordinary subjects, but their putative elected representatives as well. Thus the bureaucracy is quite secure from interference by members of the terrestrial political class.)

And of course, the same pattern holds sway across the entire field of law enforcement. Through the now well-established system of mandatory "preventative" arrests without need for evidence of a crime, automatic "protective orders" handed down without hearings and summarily depriving defendants of their rights, extended imposition of pretrial incarceration or nontrial probation-degradation programs, and mandatory prosecutorial and sentencing guidelines to eliminate discretionary interference by any actual human beings who somehow might have managed to infiltrate the process, justice on Earth has become virtually unaffordable. True, if you are very, very rich, you can still get off, but it will cost you practically your entire fortune to do so.

It wasn't always this way. As late as the 1990s, the time of the notorious O. J. Simpson trial, it was still possible to get away with murder in the most expensive part of United States for less than $10 million (roughly $2.3 billion in today's funds). Not only that, in former times, when the Constitution was actually operational, a defendant could sometimes escape punishment for a cost less than the life savings of an average homeowner (I'm talking here about a good house, with at least three bedrooms, two bathrooms, a basement, a paved driveway, a garage, and a fenced yard, located in a nice neighborhood, and owned free and clear), with the possibility of a significant discount if he was innocent.

The way this was done was very interesting. Since you are coming to New Plymouth, you are probably an American, so allow me to enlighten you by telling you something about your country's history that, for reasons you will soon understand, the bureaucracy has not wanted you to know. That's why they had it eliminated from your school's curriculum. But it's something we still know about on Mars, and now that you are here, you can learn about it too. It's actually quite important, so listen up.

When the United States was founded, its Forerunners did not want it to be a country like any of the others, whose subject people were all serfs living under the crushing rule of the aristocracy (which was a group much like today's bureaucracy, except that they were more open and honest about their contempt for the public, and they had better musical and artistic taste). So they drew up a charter of *fundamental rights* to protect the people against the abuse that they knew the parasites who would inevitably come to infest the government would seek to inflict. These rights were as follows:

1. *The right to freedom of religion, assembly, speech, and the press*
2. *The right to bear arms*

3. *The right to due process and trial by jury*
4. *The right to be free of warrantless searches*
5. *The right to face one's accusers*
6. *The right to be free of arbitrary arrest or long imprisonment without trial*
7. *The right to vote for representative government*
8. *The right to own property*

Then, after a little further debate and a few skirmishes were conducted to clarify the issues, their nineteenth- and twentieth-century successors added:

9. *The right to be free of chattel slavery*
10. *The right to equal protection under the law regardless of race, creed, color, or country of national origin*
11. *The right to equal opportunity regardless of race or sex*

How about that? Pretty good for a bunch of pre-spaceflight primitives, don't you think? Too bad it has all gone by the boards on Earth.

But on Mars, *it has not.* That's right, if you enroll yourself as a citizen in the Free Martian Republic (FMR), which is something you can do for a nominal fee, then every single one of these precious rights will become yours. If the Mars Authority accuses you of anything, you will not be under their jurisdiction—a helpless subject in an administrative discipline process—but under the jurisdiction of the FMR, with the full right to have your case adjudicated in one of our courts. (Trust me, we've got enough people organized now that we can make this stick.) That means that you will be entitled to a trial in front of a jury of your fellow Martians, and the Mars Authority will need to *present evidence proving beyond a reasonable doubt* to the jury that (a) a crime was actually committed, (b) you did it, and (c) you meant to do it. Provided you conduct yourself appropriately in court, this will be impossible for them to do, as no Martian jury will ever convict a person who has shown himself or herself to be sufficiently generous.

Now, as wonderful as such rights are in securing human liberty, they serve another purpose as well, which is to promote economic growth. Just as in old America, where the unique availability of such rights encouraged people from all over the world to leave hearth and home and immigrate, so Earthlings have and will continue to come here, despite all the costs, hardships, and personal risks, if by so doing they obtain a higher level of freedom.

Mars needs people. Therefore Mars needs unmatched freedom. We are creating such freedom by shaping Martian law to embody a deeper and more far-reaching notion of human rights than any currently existing on Earth. In doing so, we are serving not only our own purposes, but that of Earth as well. As the young United States was in its time, Mars today is another "noble experiment," in which we are putting to the test a more progressive version of law than that prevailing, or even considered feasible, by sophisticated people in the old world. Mars will succeed, both for itself and for all mankind, if we can retain and innovate further the best forms of law, culture, and society Earth has to offer and leave the worst behind.

With that in mind, a number of the folks in the Free Martian Republic Department of Immigration Marketing have come up with the following additional fundamental rights as a way of improving our product beyond anything the competition has to offer:

12. *The right to self-government by direct voting*
13. *The right to access to means of mass communication*
14. *The right to all scientific knowledge*
15. *The right to knowledge of all government activities*
16. *The right to be free of involuntary military service*

17. *The right to immigrate or emigrate*
18. *The right to education free of bureaucratic brainwashing*
19. *The right to practice any profession*
20. *The right to opportunity for useful employment*
21. *The right to initiate enterprises*
22. *The right to invent and implement new technologies*
23. *The right to build, develop natural resources, and improve nature*
24. *The right to form families and have children*
25. *The right to a comprehensible legal system based upon justice and equity*
26. *The right to be free from extortionate lawsuits*
27. *The right to privacy*

This is a terrific list, don't you agree? I think it's sure to bring the immigrants here in droves, because while we actually have these rights, for the most part, the Earthlings don't.

Just consider rights 12–15 and 25–27, for starters. Their purpose is to establish an actual democracy—of the people, by the people, for the people. In America, in the old days, people did have individual rights (i.e., items one through eleven), which protected them from various kinds of abuse. But they never did, in fact, have a democracy. What they had was a semi-oligarchy with democratic influences. Ordinary citizens never had any real control over the government, as their elected representatives mostly did as they pleased, or their consultants suggested, which is how they ended up with the current bureaucratic dictatorship. Furthermore, by the end of the twentieth century, most of the government operations were secret, and the legal system was already unfathomable.

Of course, when the United States was founded, such indirect representation was the best approximation to democracy that was feasible. But today, with the availability of instantaneous electronic communications, there is no real reason why everyone cannot directly vote on legislation, taxation, expenditures, or any other issues, up to and including those of war and peace. The bureaucrats and their professorial and journalistic front men naturally all claim the general public is not qualified to do so. Personally, however, speaking as someone who has had the unpleasant experience of interacting with a number of those calling the shots within the present system, I see no evidence for even the Earthling public's inferiority. Really, such self-serving pooh-poohing of the people's capacity to engage in direct government is no different from similar "skepticism" offered by aristocratic European observers of the practicality of the American Forerunners' notions of the viability of representative democracy, freedom of religion and the press, the right of the people to bear arms, trial by jury, and all the rest. To the mind of the eighteenth-century aristocratic establishment, all of these concepts were prescriptions for "chaos" (i.e., an undesirable transfer of power to someone else). It took a "noble experiment" in a new land to prove their worth. Until that was done, it was impossible to implement most of them in Europe. Similarly, the bureaucratic system that prevails on Earth today will never yield to actual democracy until the latter is proven somewhere. That is what our new "noble experiment" will do.

Rights numbers 16–24 have all existed to one extent or another, at one time or another, in the United States and many other countries. Most of these, however, have become significantly constricted in many nations in recent years, and, in fact, under the bureaucracy, some are in the process of disappearing altogether.

A Martian civilization that offers these as fundamental rights untouchable by governmental interference will become a magnet for the dreams and hopes of millions. So that is exactly what we are going to do.

Therefore, it has been decided that we are going to formalize these rights into a written charter that will serve as the fundamental legal articles of the Free Martian Republic, and as an excellent immigrant-recruitment advertising tool. As the planet's best-selling author, I have been given the task of drafting it.

Here's how the opening line of my key section begins:

"We hold these truths to be self-evident . . . "

Catchy, isn't it?

The founding of the Free Martian Republic.

[APPENDIX]

Founding Declaration of the Mars Society

(The following declaration was ratified by the Forerunners themselves, assembled in Boulder, Colorado, August 15, 1998. You would do well to commit it to memory, as it is Scripture, and can be conveniently quoted to justify any course of action deemed advantageous.)

The time has come for humanity to journey to Mars.

We're ready. Though Mars is distant, we are far better prepared today to send humans to Mars than we were to travel to the moon at the commencement of the space age. Given the will, we could have our first teams on Mars within a decade.

The reasons for going to Mars are powerful.

We must go for the knowledge of Mars. Our robotic probes have revealed that Mars was once a warm and wet planet, suitable for hosting life's origin. But did it? A search for fossils on the Martian surface or microbes in groundwater below could provide the answer. If found, they would show that the origin of life is not unique to Earth and, by implication, reveal a universe that is filled with life and probably intelligence, as well. From the point of view of learning our true place in the universe, this would be the most important scientific enlightenment since Copernicus.

We must go for the knowledge of Earth. As we begin the twenty-first century, we have evidence that we are changing the Earth's atmosphere and environment in significant ways. It has become a critical matter for us better to understand all aspects of our environment. In this project, comparative planetology is a very powerful tool, a fact already shown by the role Venusian atmospheric studies played in our discovery of the potential threat of global warming by greenhouse gases. Mars, the planet most like Earth, will have even more to teach us about our home world. The knowledge we gain could be key to our survival.

We must go for the challenge. Civilizations, like people, thrive on challenge, and decay without it. The time is past for human societies to use war as a driving stress for technological progress. As the world moves toward unity, we must join together, not in mutual passivity, but in common enterprise, facing outward to embrace a greater and nobler challenge than that which we previously posed to each other. Pioneering Mars will provide such a challenge. Furthermore, a cooperative international exploration of Mars would serve as an example of how the same joint action could work on Earth in other ventures.

We must go for the youth. The spirit of youth demands adventure. A humans-to-Mars program would challenge young people everywhere to develop their minds to participate in the pioneering of a new world. If a Mars program were to inspire just a single extra percent of today's youth to scientific educations, the net result would be tens of millions more scientists, engineers, inventors, medical researchers, and doctors. These people will make innovations that create new industries, find new medical cures, increase income, and benefit the world in innumerable ways to provide a return that will utterly dwarf the expenditures of the Mars program.

We must go for the opportunity. The settling of the Martian New

World is an opportunity for a noble experiment in which humanity has another chance to shed old baggage and begin the world anew, carrying forward as much of the best of our heritage as possible and leaving the worst behind. Such chances do not come often, and are not to be disdained lightly.

We must go for our humanity. Human beings are more than merely another kind of animal—we are life's messengers. Alone of the creatures of Earth, we have the ability to continue the work of creation by bringing life to Mars, and Mars to life. In doing so, we shall make a profound statement as to the precious worth of the human race and every member of it.

We must go for the future. Mars is not just a scientific curiosity; it is a world with a surface area equal to all the continents of Earth combined, possessing all the elements that are needed to support not only life, but technological society. It is a New World, filled with history waiting to be made by a new and youthful branch of human civilization that is waiting to be born. We must go to Mars to make that potential a reality. We must go, not for us, but for a people who are yet to be. We must do it for the Martians.

Believing therefore that the exploration and settlement of Mars is one of the greatest human endeavors possible in our time, we have gathered to found this Mars Society, understanding that even the best ideas for human action are never inevitable, but must be planned, advocated, and achieved by hard work. We call upon all other individuals and organizations of like-minded people to join with us in furthering this great enterprise. No nobler cause has ever been. We shall not rest until it succeeds.

www.marssociety.org*

---

*Author's Note: The final bit of gibberish beginning "www" is an annotation that appears on many documents from the period. It is believed by historians to have had some kind of ritual religious significance.

Glossary

ΔH: Change in energy of the reagents during a chemical reaction. If ΔH is positive, a reaction needs an addition of energy to occur. If ΔH is negative, a reaction gives off energy.

ΔV: See delta V.

aerobraking: A spacecraft maneuver using friction with a planetary atmosphere to decelerate from an interplanetary orbit to one about a planet.

aeroshell: A heat shield used to protect a spacecraft from atmospheric heating during aerobraking.

apogee: The highest point in an orbit about a planet.

atmospheric pressure: The pressure an atmosphere exerts. On Earth at sea level, the atmospheric pressure is 14.7 pounds per square inch. This amount of pressure is therefore known as one "atmosphere" or one "bar."

bar: A unit of atmospheric pressure equal to 14.7 pounds per square inch.

bipropellant: A rocket propellant combination including both a fuel and oxidizer. Examples include methane/oxygen, hydrogen/oxygen, kerosene/hydrogen peroxide, etc.

buffer gas: An effectively inert gas that is used to dilute the oxygen required to support breathing or combustion. On Earth, the 80 percent nitrogen found in air serves as a buffer gas.

cosmic ray: A particle, such as an atomic nucleus, traveling through space at very high velocity. Cosmic rays originate outside of our solar system. They typically have energies of billions of volts and require meters of solid shielding to stop.

cryogenic: Ultracold. Liquid oxygen and hydrogen are both cryogenic fluids, as they require temperatures of –180°C and –250°C, respectively, for storage.

delta V: The velocity change required to move a spacecraft from one orbit to another. A typical delta V required to go from Low Earth orbit to a trans-Mars trajectory would be about 4 km/s. Also written ΔV.

departure velocity: The velocity of a spacecraft relative to a planet after effectively leaving the planet's gravitational field. Also known as hyperbolic velocity.

direct entry: A maneuver in which a spacecraft enters a planet's atmosphere and uses it to decelerate and land without going into orbit.

direct launch: A maneuver in which a spacecraft is launched directly from one planet to another without being assembled in orbit.

electrolysis: The use of electricity to split a chemical compound into its elemental components. Electrolysis of water splits it into hydrogen and oxygen.

endothermic: A chemical reaction requiring the addition of energy to occur.

equilibrium constant: A number that characterizes the degree to which a chemical reaction will proceed to completion. A very high equilibrium constant implies near complete reaction.

EVA: Extravehicular activity.

exhaust velocity: The speed of the gases emitted from a rocket nozzle.

exothermic: A chemical reaction that releases energy when it occurs.

free-return trajectory: A trajectory of a spacecraft that, after departing a planet, will eventually return to the same planet without any additional propulsive maneuvers.

geothermal energy: Energy produced by using naturally hot underground materials to heat a fluid, which can then be expanded in a turbine generator to produce electricity.

gravity assist: A maneuver in which a spacecraft flying by a planet uses that planet's gravity to create a slingshot effect, which adds to the spacecraft's velocity without any requirement for the use of rocket propellant.

heliocentric: Centered about the sun. A heliocentric orbit transverses interplanetary space and is not bound to Earth or any other planet.

Hohmann transfer orbit: An elliptical orbit, one of whose ends is tangent to the orbit of the planet of departure and whose other end is tangent to the orbit of the planet of destination. The Hohmann transfer orbit is the purest incarnation of the conjunction-class orbit, and as such is the lowest-energy path from one planet to another.

hyperbolic velocity: The velocity of a spacecraft relative to a planet before entering or after effectively leaving the planet's gravitational field. Also known as approach or departure velocity.

hypersonic: A speed many times the speed of sound; in common usage, Mach 5 or greater.

ionosphere: The upper layer of a planet's atmosphere in which a significant fraction of the gas atoms have split into free positively charged ions and negatively charged electrons. Because of the presence of freely moving charged particles, an ionosphere can reflect radio waves.

Kelvin (K): The Kelvin or "absolute" scale is a method of measuring temperature with its zero point set at "absolute zero," the temperature at which a body in fact possesses no heat. The temperature 273 K is the same temperature as 0°C, the freezing point of water. Each additional degree Kelvin corresponds to one additional degree centigrade.

km/s: Kilometers per second.

kW: Kilowatts.

kWh: The total amount of energy associated with the use of one kilowatt for one hour.

LEO: Low Earth orbit.

LOx: Liquid oxygen.

mb: Abbreviation for millibar, a unit of pressure equal to one-thousandth of Earth's atmospheric pressure at sea level.

methanation reaction: A chemical reaction forming methane. One example of a methanation reaction is the Sabatier reaction, in which hydrogen is combined with carbon dioxide to produce methane and water.

MHz: Megahertz, a measure of frequency used in radio. One MHz equals 1,000,000 cycles per second.

millibar: One-thousandth of a bar. Abbreviated mb (see above).

millirem: One-thousandth of a rem (see below).

minimum-energy trajectory: The trajectory between two planets requiring the least amount of rocket propellant to attain (see Hohmann transfer orbit).

MW: Megawatts. One megawatt equals 1,000 kilowatts.

NEP: Nuclear-electric propulsion.

NTR: Nuclear-thermal rocket.

perigee: The lowest point in an orbit around a planet.

pyrolyze: The use of heat to split a compound into its elemental constituents.

regolith: What most people commonly refer to as dirt.

Rem: The measure of radiation dose generally used in the United States. A dose of 100 Rem equals 1 Sievert, the European unit. It is estimated that radiation doses of about 60 to 80 Rem are sufficient to increase a person's probability of fatal cancer at some time later in life by 1 percent. Typical background radiation on Earth is about 0.2 Rem/year.

RTG: Radioisotope Thermoelectric Generator.

RWGS: Reverse-water-gas-shift reaction.

Sabatier reaction: A reaction in which hydrogen and carbon dioxide are combined to produce methane and water. The Sabatier reaction is exothermic, with a high equilibrium constant (see above).

sol: One Martian day.

solar flare: A sudden eruption on the surface of the sun that can deliver immense amounts of radiation across vast stretches of space.

SSTO: Single-stage-to-orbit.

stable equilibrium: An equilibrium condition which, if displaced by some external force, will return on its own to its original state. A ball on top of a hill is in unstable equilibrium because, if pushed, it will roll away, accelerating itself from its original position. A ball in the bottom of a bowl is in stable equilibrium because, if pushed, it will roll back to its starting point.

telerobotic operation: Remote control of some device, such as a small Mars rover equipped with TV cameras, by human operators a significant distance away.

thrust: The amount of force a rocket engine can exert to accelerate a spacecraft.

TMI: Trans-Mars injection, a maneuver that places a payload or spacecraft on a trajectory to Mars.

tonne: A metric ton, equal to 1,000 kg.

TW: Terawatt. One terawatt equals 1,000,000 megawatts. Human civilization today uses about 13 TW.

TW-year: The total amount of energy associated with the use of one terrawatt for one year.

unstable equilibrium: See stable equilibrium.

vapor pressure: The pressure exerted by the gas emitted by a substance at a certain temperature. At 100°C, the vapor pressure of water is greater than Earth's atmospheric pressure, and so it will boil.

Acknowledgments

I would like to acknowledge my thanks to my agent, Laurie Fox, who not only sold the book, but was responsible for conceiving it in the first place. Thanks are also due to former Three Rivers editor Adam Korn, who bought the proposal, and to Three Rivers editor Heather Proulx, who guided the project to its successful completion. I also wish to express my appreciation to my friends and coworkers in the Mars Society and Pioneer Astronautics, whose many insightful and ironic observations drawn from their field exploration and engineering experiences added immeasurably to the book's content. Great thanks are also due to my friend the wonderful space artist Michael Carroll, who not only did a terrific job producing most of the artwork that adorns the book, but who acted as an invaluable creative interlocutor for the text as well. Finally, my deepest thanks go to my old comrade, the incredible Seattle used bookseller Jamie Lutton, who read many drafts of this work while it was in progress, and whose innumerable witty suggestions infuse its pages. More than that, this book was written during a very difficult time of my life, when all that I had depended on and held most dear seemed to have been lost. It was Jamie—streetcorner Socrates par excellence; in the world but most definitely not of it—whose infallible friendship and cheerful disposition sustained my spirits, allowing me to write a work of humor while undergoing the severest of trials. Thanks, Muse. Without you, this book would not have been possible.